THE
SOURCE

THE
SOURCE

The Secrets of the Universe, the Science of the Brain

DR. TARA SWART

HarperOne
An Imprint of HarperCollinsPublishers

HarperOne

HarperCollins books may be purchased for educational, business, or sales promotional use. For information, please email the Special Markets Department at SPsales@harpercollins.com.

First published in 2019 by Ebury. Ebury is part of the Penguin Random House group of companies.

FIRST US EDITION

Illustrations © Ollie Mann

Library of Congress Cataloging-in-Publication Data has been applied for.

ISBN 978-0-06-293573-1

19 20 21 22 23   LSC   10 9 8 7 6 5 4 3 2 1

To Robin—husband, best friend, twin flame.
And to Tom—my son and unexpected surprise!

"Some men seem to attract success, power, wealth, attainment with very little conscious effort; others conquer with great difficulty; still others fail altogether to reach their ambitions, desires and ideals. Why is this so? The cause cannot be physical . . . hence mind must be the creative force, must constitute the sole difference between men. It is *mind* which overcomes environment and every other obstacle . . ."

Charles Haanel, The Master Key System *(1919)*

Contents

Preface

A Return to The Source

"And yet the menace of the years; Finds, and shall find, me unafraid . . . I am the master of my fate; I am the captain of my soul."

Invictus, William Ernest Henley

Life-changing opportunities pass us by every day. A chance encounter with a potential partner, a comment that could open up a new career; in our frenetic lives such fleeting moments are easily overlooked or dismissed. But with the right help we can train our minds to notice them and to seize them—and learn how to transform a split-second into a lasting change. That's because the things we most want from life—health, happiness, wealth, love—are governed by our ability to think, feel and act; in other words, by our brains.

This has long been the promise of New Age thinking: that we can control our destiny by reshaping our minds. Despite skepticism, systems like *The Secret* and its forbears have been used by millions of people for generations. Why? Because, if we strip away the mystique, some of the concepts underlying them are fundamentally powerful. Even more importantly, they are now being backed up by the latest breakthroughs in neuroscience and behavioral psychology.

I am a medical doctor specializing in psychiatry, a neuroscientist turned executive coach and a speaker on mental performance. All my research and practice has led me to be convinced of one thing: our ability to actively alter the way that our minds work. Science and evidence-based psychology are proving that we all have the power to manifest the lives we desire.

This is personal: I have been on my own journey to understanding that the key to creating the life I dreamt of lay within my brain—what I call "The Source." In this book, I will share the thinking and techniques that have worked for me and my clients, as well as cutting-edge psychological research and neuroscience—all distilled into in a rigorous, proven toolkit for unlocking your mind. If you are open to change, and follow these simple steps, then you will be ready to harness more of those life-changing opportunities and reach your fullest potential.

As the eldest child of first-generation immigrant Indian parents, I was brought up in north-west London in the 1970s and 1980s with a curious mixture of cultural beliefs, foods and languages. I quickly learned to adapt to the various ideas swirling around me, but deep down I felt conflicted. At home, yoga and meditation were part of the daily routine, we adhered to a strict vegetarian, Ayurvedic diet (in which turmeric was lauded as a cure-for-all) and we had to offer the food to God before taking a mouthful. My parents tried to explain the benefits of these practices but they sounded far-fetched and I just wanted to fit in with my friends. I dreamt of a simple life where what went on at home was in harmony with the world around me.

At school, my younger brother and I would never mention the idea of reincarnation, but at home we had to worship our ancestors with prayers chanted over incense sticks and with food offerings, sure in the knowledge that our ancestors were walking around in another dimension and able to exert an influence on our lives. Much to my amazement, it was casually accepted in my extended family that I was the reincarnation of my late grandmother. Her purported regret at not having had a formal education (she grew up in a village in India) led to a weight of expectation on me to become a medical doctor—the ultimate respected profession in my culture, akin to godliness because of its association with life and death. I dutifully went through school and to university on this path that had been laid out for me without question.

I believe I was drawn to psychiatry and neuroscience during my first few years at college in a quest to understand myself—who I really was and what my true purpose would be if I had been free to decide. In my twenties, I rejected my cultural heritage in a bid to relieve myself of the massive expectations placed on me in childhood. I moved out of home and in with my university friends, became interested in fashion and expressing myself through my clothing, started to travel around Europe and then to South Africa, and ventured into the realm of having a boyfriend. I met my husband-to-be when we were both working in psychiatry, and we moved to Australia then later lived in Bermuda, all of which expanded my world view and understanding of people and cultures. But the real crunch point came in my mid-thirties when I suffered major personal and professional crises at the same time.

I had become increasingly unhappy in my work as a psychiatrist, worn down by the long hours and workload and the sense of not being able to make a real difference to my patients. I witnessed so much human suffering and saw how tough and cruel life was for the mentally vulnerable. I cared deeply about my patients, but I had a nagging sense that they deserved more than just medication and hospitalization—that a healthier regime and a sense of well-being could be cumulative in terms of their recovery. There was something negative for me about focusing only on illness, with at best a return to normality as the end goal for treatment. I knew that there was a world out there where a good outcome could be even better and I felt that I would contribute more to the world if I could work in a way that sought to optimize health, rather than simply treating acute symptoms after the event. In the end, I decided to leave and try to do something about it. At the same time my marriage fell apart with a disastrous impact on my own sense of identity and confidence. I felt like I was drowning, with nothing to hold onto and no end in sight. I needed to master mental resilience for my own sake as well as others'.

I was struggling with my sense of self, what I was going to do with my future and trying to understand what had led to the breakdown of my marriage. I had to learn who I was outside of the partnership that had been the foundation of my formative adult years. I experienced lows that I could not describe in words, but only in the primal howls of despair and loss. However, rock bottom gave me new clarity—from that moment emerged a determination I had not known I possessed, and a sense of being on a journey that I must progress on my own to really fulfill my potential.

Some years earlier, when I had been in a good place, I had come across the concepts of positive thinking and visualization. I was still a doctor, about 30 years old, on a round-the-world trip and happily married. I felt pretty carefree. I read a lot of personal development books because I was interested in Buddhism and Jungian psychology. This was seen as pretty "alternative" by my medically trained peers who were mostly dismissive of self-help books, but I believed there was a time and place for all sorts of ideologies. I read one book known as the "positive thinking bible," which was more esoteric: it was called *The Master Key System*, written by Charles F. Haanel and originally published in 1916. It combined beliefs such as the "law of attraction" alongside the power of visualization and meditation. I didn't do any of the step-by-step exercises in the book, but it resonated with me and I resolved to return to it later "*if* I ever needed to." I forgot about it for years. It wasn't until the demolition of my lovingly built life—the end of my marriage, a massive career change, living alone and starting a new business—that I found myself drawn to it once again.

I was astounded by the power of the exercises within the book. Each week led to new and deeper insights into the thought patterns that had allowed my life to spiral out of control, and the knowledge that mastering my emotions to make them work for me rather than allowing them to overwhelm me was the answer. It was a choice! The visualizations, in particular, were a powerful revelation in comparison to just overanalyzing every situation. I moved from a sense of drowning to creating an image of a little life raft I could cling onto for as long as needed. Ahead of me was an island and I had a vision of standing on the golden sand, water running off me, the sun warming me, and one day being safe and strong. Also, I began the habit of making annual action boards. As you will see later on in the book, action boards are collages created to symbolize goals and aspirations, and keep you motivated to achieve them. My boards started with small wins, but ended up becoming something more powerful than I could ever have wished for. I've still got them from years ago—it's astonishing how much of what was on them has come true for me, right down to the small details.

I felt a great sense of power over my own life as well as the impact I could have on other people, and even the world, simultaneously following my passions and educating people. I had always believed in the importance of abundance and generosity—these were strong intrinsic values for me—but the twist was that putting that knowledge out into the world—be it to friends, family or patients—became more important than striving for what I needed to survive. I had seen myself evolve and change fundamentally, and, most crucially, my learnings came about in a manner that was fascinating to decode through my knowledge of neuroscience and psychiatry—I was able to make my own brain work differently to support my new way of living.

As I battled to forge a new path, switching my focus towards the science of brain optimization, I came to fully understand how we can harness the power of the brain to direct the life we desire. I built and refined my interpretation and teachings as I grew steadily more determined to devote my life's work to unveiling the secrets of the brain.

Neuroscience has developed massively during my adult life, largely due to the advent of scanning technology. These developments gave credence to things that I had intuitively felt but hesitated to trust, as well as to the ancient wisdom from my background that did not previously mesh with modern Western life. Brain scanning has transformed our view of the influence of the brain and the importance of mental health and strength, as well as unlocking the tremendous power of brain plasticity and what it means for all of us.

With the deep parallels to my own growth and emotional metamorphosis, I was able to inextricably link who I am with what I do, and I believed that I could assist others to achieve this too. It hit me that I now understood the reason that I was on this planet. To actualize this in the real world, all I had to do was work on what I was good at, find the positives in my life—rather than focusing on what was wrong—and then this would have an impact on all that I touched.

My understanding of neuroscience and my quest to explore esoteric philosophy began to feel not like schizoid poles of thought, but like two core elements drawn towards one another to create a new charge. The experiences of my friends, family, patients and coaching clients all fed into my ideas, and again and again demonstrated the truth at the heart of my teachings.

I was finally doing work I really wanted to do and this had a rewarding influence everywhere I went. I surrounded myself with people that I knew supported me and what I was trying to achieve. I made an action board every year without fail. I kept a journal to track my thoughts and emotional responses to everything that happened, slowly counseling myself towards trust and gratitude rather than blame or avoiding true responsibility. I was fully independent and could rely on myself to draw on my internal resources, even in situations I previously thought I could not withstand. I experienced cumulative insights, fusing what I had learned about Eastern philosophy and cognitive science together into a new manifesto for living. Over time, these changes have led to immense leaps in the scope and scale of my work, and to finding personal fulfillment and security in a way that I never thought possible.

This is what I offer you here in an up-to-date, scientifically backed, secular way: my learnings and experiences together with the science; my way of retraining the brain to direct our actions and emotions to manifest our deepest dreams. Understanding and taking control of our own brain is the key, and this is the power of The Source. It has taken me nine years of college, seven years of practicing psychiatry and ten years of being an executive coach to get to this point. But I will share with you, in this book, the secret of mastering your brain to transform your life.

Dr. Tara Swart, London

Introduction

"Whether you think you can or you think you can't—you're right."

Henry Ford

Once upon a time, we humans roamed the surface of the earth alongside other primates and even larger, stronger and more agile animals. We were no more special or extraordinary than any other creature that existed on this planet. We had a smaller skull than we do now that largely consisted of the limbic brain—the deep, ancient, emotional, intuitive part of the brain—surrounded by a slim sliver of outer cortex. And then . . . we discovered fire.[1]

We do not know if we naturally evolved to develop a larger cortex and with this came the capacity to utilize tools and make fire; or if it started with a random small spark that would change the world. But we nurtured it, not just to stay warm, but also to cook meat and therefore digest protein more efficiently. Our digestive tract shrank and we diverted resources to grow the outer cortex of the brain, which eventually became as dense as the limbic brain it surrounds. This rapid development of the cortex was the single most pivotal moment in the cognitive evolution of humankind. And it has made us by far the most successful animal on this planet.

With the growth of the modern, rational part of the brain came articulated speech and the ability to predict and plan for the future. As we became more logical and able to communicate and exist in larger tribes, we talked more and *felt* less.

We moved away from emotions to logic and facts, and survival through competition became our means to an end. We lost the sense of abundance that had got us so far, and the sense that there was enough for everyone. We lost our relationship with fate: we needed to *control* fate and *have* more than everyone else. We stopped the simple life connections—like sitting around a campfire telling stories, gazing at the stars or walking barefoot in nature—to cultivate agriculture and create industries where power and status outweighed collaboration or peaceful coexistence. We stopped *being* and started *doing* a lot of things, existing on a kind of autopilot that we could not turn off.

Millennia later, we live in a world where logic is massively overrated, emotions are seen as a weakness and decisions based on intuition have little or no place. We have forgotten where we came from. Over time, we have neglected the limbic brain that got us to the pivotal moment in our evolution, and instead placed the cortex on a pedestal. We have demoted depth, passion and instinct and come to rely on the surface-level capabilities—such as exams, rote-learning or transactional relationships—that are more connected with material gain than true joy. We live a life dominated by stress and are too busy to really take notice of who we are, where we are going and what we want from life. We are now at a moment where technology will disrupt our minds and bodies more than we can begin to imagine. We are on the threshold of massive change.

Faith in Science

What is happening to our brain now, in the midst of all this change? When I was growing up, brain scanners didn't even exist. Now, through sophisticated imaging of healthy brains, we can really see what thoughts "look like," and how anger, sadness and joy appear in the brain. Through scanning and other research, we can evidence the impact of parenting behaviors and relationships on children's brains. In adults, too, we now understand that everything, from physical exercise and meditation to social relationships and stress, is molding and shaping our brains constantly. This gives us a new context in which to make sense of age-old ideas.

Until now, the best-known proponents of the idea that we can create the life we want by changing the way we think have met with universal criticism from scientists for their suggestion that thoughts themselves are "magnetic," with a vibrational frequency that moves outwards into the universe and is met with an effect. These claims for "vibration" and "resonance" have not been grounded in empirical science. As a result, subscribing to them has, until now, required a leap of faith: asking us to choose to believe that if we think positively, we will attract many desired elements into our life. This can create an impression of a passive process where we could sit at home or on a desert island and miraculously change the world. Of course, this is impossible and, to the skeptic's ear, advocates of the "law of attraction," "manifesting" and "abundance" make the whole thing sound like a manifesto for magical thinking.

However, just as modern science shows us how age-old practices such as mindfulness and many of the ancient medical principles of Ayurveda have demonstrable, evidence-based benefits, so our understanding of neuroplasticity (the brain's ability to flex and change) demonstrates that directing our thoughts can influence not only our perception of "reality" but also our material life circumstances, our relationships, and the situations we attract into or tolerate in our lives. The way we think determines our life. This is a simple idea, but a powerful one.

Our brains actively grow and change during childhood. In contrast, as adults, we have to *consciously* direct ourselves to grow and develop as people. Quite how much we can use the inherent flexibility of the brain to enhance our experience of life is actually mind-blowing, and this concept will underpin all the theory and practice outlined in *The Source*.

If you're already a follower of the law of attraction and/or if you've read the foundational texts on the subject, you likely had to make a "leap of faith" in order to begin your journey. Perhaps this lack of supporting science made you suspicious and kept you from accessing this life-changing philosophy, or prevented your family and friends from believing in your new lifestyle. In contrast, *The Source* brings together rigorous science, studies and research to offer you a path you can believe in and share with even the most skeptical loved one. As a neuroscientist and a medical doctor with extensive experience practicing psychiatry, I make these claims with informed authority. My approach is strongly based on the proven brain–body connection—the concept that the brain and body are inextricably linked and have knock-on effects on each other, mostly via the neuroendocrine system (which includes all our glands and hormones) and the autonomic nervous system (which is all our nerves apart from the brain itself and the spinal cord). Our ability to thrive is governed by the physical condition of our (emotional and logical) brain and the quality of thoughts that we allow to arise from it. Whatever that condition and quality may be today, the elasticity of our brain means that we have the ability to change our brain pathways, and therefore our lives, for a better tomorrow.

We will at times have to challenge the consequences of our evolutionary "hardwiring" and retrain ourselves to think in a more agile and positive way. But now's the time to take action. This is not blind faith, but faith based in science.

Our Brains on Autopilot

Let's start by thinking about how we regard our brains. The first step to unleashing the power of our brain is to stop taking it for granted—our brain is our biggest asset; crucially, it governs our life: our confidence, relationships, creativity, self-esteem, life purpose, resilience, and so much more.

In every millisecond, our brain's 86 billion neurons (brain cells) are interpreting and responding to the glut of sensory feedback they receive from our body and environment, and processing and filing this information according to what the brain thinks it "means." These neurons are firing off constantly, making connections and building "pathways" as they bring together emotions and actions and memories and connections.

This is largely at a subconscious level, and we act all the time based on this "feedback loop" as it is called, adjusting and readjusting our response in real time to the perceived trigger. Information comes in from the outside world and we respond based on pattern recognition that becomes more and more ingrained as we mature and we become more stuck or set in our ways. From taking the same route to work every day (mostly without even registering the view from the window or the roads we walk down) to following familiar (and unsatisfactory) relationship models, we live huge parts of our lives on this "autopilot." We do this unquestioningly for the most part, as, through repetition, this becomes the default pathway in our brain; the longer we have been doing something, the less likely we are to question it, whether that is our favorite color or our choice of life partner.

Following this autopilot means that life follows very familiar patterns, which is much more efficient for the brain as it requires less energy. Also, the brain is wired to avoid change which it perceives as a "threat" and to which it creates a stress response that stops us from taking risks and powers down our higher thinking (the executive functions of the brain, such as regulating emotions, overriding biases, solving complex problems and thinking flexibly and creatively) to keep us safe. It opts for instant gratification and the path of least resistance whether or not it is in our best interest for it to do so. On autopilot, we don't question where these underlying, entrenched habits come from and whether they serve us any longer; we switch off and let life happen to us, assuming that much of it is out of our control. But every single thing that we do reinforces a pattern or pathway, and consolidates our autopilot behaviors. In doing so, the underlying concept that things are just the way they are convinces us more than ever that life happens to us and we are to a large extent powerless to control it. However, neuroscience shows us that we can take back control of our minds by rewiring our brain's pathways to make lasting, positive changes to our lives.

The Source

The Source is the incredible, complex and sophisticated thing that is our *whole* brain—not just the cortex and our planning and data-driven decision-making abilities. The true power of the brain lies in being able to integrate what we think with how we feel—the cortex *and* the limbic system together—with what our gut tells us and what we sense throughout our entire bodies. This creates an experience of life of which we can take true ownership; one that is filled with a trust in our own amazing ability to navigate circumstances with every part of us aligned and fully immersed.

Life does not have to be about fear and half-measures or what-ifs and regrets. We each have the capacity in our brains to live life fully, boldly and without shame or sadness. I have come to learn, through the combination of my cultural heritage with modern medicine and neuroscience, that if we access the full potential of our brains, we can live life very, very differently than the way we have before today.

At the heart of The Source is developing a level of awareness about our neural pathways and the patterns in their activity that dictate how we unconsciously react to triggers and events— like losing our temper versus shutting down, comfort eating or reaching out for help. Becoming more aware of our responses and behavior can help us shape our reactions to the challenges we encounter in our lives. This awareness of our own mental state and that of others governs our most complex and critical social interactions: to make sense of the cognition of others is known as "theory of mind" and we use this to interpret, understand and predict the actions of those around us.

There are obvious advantages to being skilled at understanding what is motivating others' actions, and an extreme example of not being able to do this is seen in the autism spectrum. We can become more skilled at doing these things using the power of neuroplasticity (meaning we can learn and change) and brain agility to use whole-brain thinking in a variety of situations and with different types of people.

Developing metacognition, or "thinking about thinking," and becoming "aware of one's awareness," rather than functioning on autopilot is one of the main objectives of The Source. A function of the pre-frontal cortex (PFC), the term comes from the root word *meta*, meaning "beyond." The PFC monitors sensory signals from other regions and uses feedback loops to direct our thinking by constantly updating our brain depending on what is playing out in the outside world. Metacognition encompasses all of our memory-monitoring and self-regulation, consciousness and self-awareness—crucial capacities to regulate our own thinking and maximize our awareness and our potential to learn and change.

My four-step program to fully awaken the brain and own your fullest potential is the product of a long-standing neuroscientific and psychiatric understanding of the brain, updated with cutting-edge cognitive science. I combine this with a healthy measure of esoteric- and spirituality-based supports. We will explore, in detail, the neuroscience behind your ability to "create" your future—the law of attraction and how you can train your brain to "manifest" your dreams. We will look at the power of visualization informed by twenty-first century science and why it works, as well as what happens inside the brain when you proactively channel positive thoughts. We will delve into how using action boards to focus your intentions will allow you to construct a life that is true to your innermost needs and aspirations. The result is a potent cocktail that will reboot your thinking and motivation, and halt negativity in its tracks.

The Source and you

This book is a guide to life that combines science and spirituality in a way that is open-minded and practical. I want you to be able to wake up your brain and unlock its full potential so that you can make of life what you really want, turn off that autopilot and make the decisions necessary to move forward. We all suffer from negative thought patterns and behaviors, from habits that make life easier but aren't helpful to our happiness, and from emotions that limit our life choices; but if they are determining how you live, then you need to find a way past them. This is your chance to fully engage with your deepest needs and desires, and to make life happen *for* you, not *to* you.

Take a look at the statements overleaf and see how many resonate with you. If you find yourself nodding your head, this book is for you.

Relationships

- I care more about everyone else than I do about myself.

- I struggle to find and develop healthy relationships and suspect there is a pattern to the problems that keep cropping up that threatens my long-term happiness.

- I've been so badly hurt in the past that I have totally shut off from ever meeting anyone special.

- I'd settle for any kind of relationship rather than be single or in the dating game.

- I feel desperate to meet someone so I'm not the last childless person in my group of friends.

- I feel stuck in an unhappy long-term relationship and can't see a way out.

- I am unable to make new friends, and don't feel connected to my established friendship groups. I don't know how to "move on."

- I want a partner and a family, but don't feel I have any control over this happening.

Work

- I agonize over what decisions to make.

- I know I can be the best at something, but I'm not sure I'm doing the right things to fulfill my potential.

- I've never asked for a pay raise or promotion.

- My work bores me but it pays the bills.

- I have a fixed view of what I can and can't do and I'm resigned to this.

- I get so tired and burnt out sometimes that I can't get out of bed.

- I have a lot of ideas of what I want from my career, but don't know how to make them happen.

Personal development

- "Never," "always" and "should/must" dominate my thinking.

- I wish I was more in control of my life.

- I feel directionless and worry that life will pass me by.

- I get overwhelmed by extremes of emotion.

- The way I feel about my body and appearance is highly influenced by what mood I am in.

- I sometimes resent people—even my friends—who have a better lifestyle than me.

- I give most people a curated version of my life because deep down I don't think it's that great.

- I'd like to start a new venture or go traveling—do something different with my life—but I keep putting it off.

If you recognize any of these statements, then understanding how neuroplasticity works will allow you to think differently and change old assumptions and deeply embedded self-beliefs. This book will enable you to set your intention, goal or dream for your future and show you how you can manifest the ideal vision of what you want your life to look like.

Journaling

Before we begin, it's time to start an important new habit: journaling. Throughout *The Source* we're going to be asking ourselves lots of questions, uncovering debilitating patterns and habits, and building steps towards a potential brighter future, so take the time to get yourself a journal to fill in that makes you feel happy and empowered.

To get the maximum benefit from this practice, you'll need to write in your journal daily about your thoughts and reactions to events and the people in your life. You don't need to write long entries, but aim to be honest and open about your emotions, motivations and behaviors.

TARA: UNCOVERING MY TRUTH

Even as my career star was rising, it was as if I was paralyzed with fear when it came to my romantic life. I avoided intimate relationships like the plague, going from non-committal dating for a few years to not dating at all for two years before I realized I was being controlled by my lack of trust, and fear of being hurt again after the breakdown of my marriage. I had absolutely convinced myself that it was the best and right thing for me to never consider getting married again, and therefore that being open to a committed relationship was a total waste of my time and energy. I had to work hard to challenge this belief—my own emotions created my biggest barrier to change.

I found my journaling a powerful way to see that mistrust was leading me into repeated patterns of holding back from or avoiding intimacy that inevitably led to self-fulfilling prophecies. I decided to act as if the past had no hold over me and just tried to think and behave differently to see if my worst fears would really come true. They didn't. There were a few bumps in the road but I also learned that that was okay and I could still move forward with trust until everything worked out. Shutting off was definitely not going to get me there.

Tapping into The Source

There is a neuro-myth (that won't go away!) that we only use 10 percent of our brains. It's not actually true, but our fondness for this statistic belies the scientific truth that the potential to grow and change our brain and how it directs our lives is far greater than we have been led to believe.

In this book, there are no wild claims or talk of quantum physics. There will always be enough rigorous science to back up the theory. I'll share the four-step plan I have used with patients and coaching clients, and also the practices that I have personally benefitted from, such as visualizations, journaling and creating powerful action boards to bring your desires to life.

In Part 1: Science and Spirituality, you'll learn about the science behind the law of attraction and the power of visualization. Part 2: The Elastic Brain explores neuroplasticity and how our brain really can change the way it works. Part 3: The Agile Brain uncovers the power of an agile and balanced brain in how we live our lives. I'll be encouraging you to try the exercises throughout the book and in the final practical section—Part 4: Fire Up The Source—to give you a roadmap to tap into The Source.

This book will take you on a journey, blending science and spirituality and helping you to turn insight into motivation and autopilot into action. This is the power of The Source—to come to understand how you are in control of your own destiny. You are just four steps away from building a new confident you, and a new magical life.

PART 1

Science and Spirituality

Chapter 1

The Law of Attraction

"Attract what you expect, reflect what you desire, become what you respect, mirror what you admire."

Unknown

Have you ever had one of those days where everything goes brilliantly: from unexpectedly having time for a leisurely breakfast because you woke up before your alarm feeling really relaxed and awake, to finding a great deal on something you've wanted to buy for ages or being offered a brilliant opportunity at work? When this happens, we say: "It must be my day," or decide we're on a "winning streak." Such opportunities appear to be random and out of our control. Or perhaps you know someone who is always "lucky." They're the person who invests in the right things at the right time, always manages to score a flight or hotel upgrade, finds and sustains a diverse and robust friend group at every turn and loves a partner who matches them well.

But I've come to understand how all of these "lucky" moments are far from purely serendipitous: they are simply the law of attraction in action. Think about particularly "lucky" or good things that have happened to you recently. A work opportunity seems like good fortune but why not consider it a reflection of your successful performance? A chance meeting with a new partner can feel more like a "golden ticket" than the result of the conscious effort you've made to be open to meeting people and being in the right place at the right time. Life is not just happening to us; we are creating it with everything that we do.

The law of attraction is at the heart of The Source. In short, it describes the way that we can create the relationships, situations and material things that come into our lives as a direct consequence of the way we think. We "manifest" them by *focusing* on them, *visualizing* them and *directing* our energy towards them through our actions. This is the idea that by choosing to focus your energy and attention towards something, it can be manifested in your life.

This idea of "manifesting" is controversial and often gets dismissed, with people writing off the whole idea of the law of attraction as a result, but I think the issue is partly a question of semantics. Books like *The Secret* and *The Master Key System* based their success on "thought vibrations" and "higher powers," so the word "manifest" became loaded with associations of religious proof and blind faith, but manifesting is merely another way of saying we "make something happen." It relates to the action rather than to mere intention. Instead of loading this word with weighty expectations of wondrous and spontaneous happenings, we should consider it as a directed and purposeful connection between intention and action. This takes the ideology of these inspirational books and backs it up with modern science.

Within the bounds of evidence-based science, I will outline six principles that underpin the law of attraction (pages 25–50), exploring the brain processes that support them, as well as practical activities that will enable you to turn these principles to your advantage, turbo-charging The Source to help you design your ideal life. Various combinations of these six principles feature in popular law of attraction manuals, and you might be surprised by how much scientific truth we can uncover in them.

Setting Your Intention

Before we consider each of the six principles in turn, I want to consider what law of attraction enthusiasts call the "intention point." They define this as the meeting place between "heart" and "mind," but science shows us that there is more than just blind faith in this notion. When we set a goal from the "intention point" what actually happens scientifically is that our intuition, our deepest emotions and our rational thinking line up and work in harmony rather than conflict. It's almost impossible to reach our goals when we are out of kilter in these three dimensions.

It's interesting how, when we're making life choices, we tend to distinguish the concepts of our head, heart and gut as separate things, often pulling us in different directions—we pit the logical processes of the brain against the more instinctive responses of our body and the pull of our emotions (in both big decisions, such as choosing a job we want versus the realities of career progression, and small choices, such as deciding whether to buy a new expensive jacket on sale or not). Increasingly, emerging science is demonstrating the interrelatedness of the relationship between our body and mind, teaching us more about the brain–body connection. Because of the two-way interaction of all our nerves and hormones, we are seeing more clearly now that if, for example, we are hungry or tired, this affects our mood and decision-making, as well as the fact that if we are depressed or stressed, we can experience changes in sleep, appetite, weight and a whole host of physical factors. By accepting this and striving for intentions that feel right at a whole-brain, and whole-body, level makes scientific sense.

PIPPA: A DIFFICULT CHOICE

When I started coaching Pippa, her marriage was breaking down. Married to a lawyer who slept at the office many nights of the week and had been a workaholic for most of the ten years they had been married, she was deeply unhappy. The couple had two small children and Pippa found herself on her own with them most weekends as her husband would be at the office or on work-related trips. She had thought about leaving him many times, and told me it felt like a "sham marriage" and that she was deeply lonely. Pippa's family and her in-laws urged her to stick with it, telling her things would get better when her husband made his next promotion and the children were older. On a practical level, Pippa knew she would not be able to afford to stay in the family home if they split up and worried about the huge life changes this would bring for her and the children.

I encouraged Pippa to create an action board using images to help her visualize her life as she wanted it to be. This proved to be a turning point for Pippa. When she returned the next week, she had made a powerful and strong board. At its center was an image of a woman with her back to the camera, at the foot of a mountain, on her own. Her hands were on her hips and she looked ready to climb. She knew she had an arduous journey ahead of her but she was confident she could do it. There were personal pictures of her children too, and pictures of places she wanted to go, and family adventures she wanted to have.

The board helped Pippa to see what it was she really wanted in her heart and gut, and identify the logical issues clouding her decision. Even though their circumstances would be diminished in the short-term, she felt empowered to make a choice that would be better for all of them in the end. She had been letting fear rule her decisions, thinking of the worst possible scenarios and catastrophizing—her anxious and fearful brain had been drowning out her emotional brain with logical issues and limitations. Through the action board and our work together she was able to align her head and heart and access her true intention for the future.

She told her husband that night that she wanted to separate. After the initial shock, he agreed. Four years on, she is happily divorced and has a good relationship with her ex-husband. She has no doubt she made the right decision.

Attracting what you really want

There is nothing mystical or magical about setting your intention. It's just a question of asking yourself: "Is my life panning out in the way I want it to?" and if the answer is no, envisaging the way you'd like it to be and taking action. It is through unlocking the full power of our brain—The Source—that we begin to think and behave in ways that will help us realize this vision.

What is key to remember is that our intention and focus are at their strongest when our goals align with our deepest life choices and values. For example, if you force yourself to make career choices for money at the expense of a sincere desire to have a sense of purpose and help other people with the work you do, you will suffer negative symptoms, such as stress and anxiety, in your physical, emotional and spiritual life as you try to live out of sync with your true self. Or if you compromise on what you need from a partner because your biological clock is ticking, you may end up finding that you have ticked some of your boxes but still have unmet needs deep down.

Such symptoms represent an internal cry of: "This isn't what I want!" These internal conflicts impact on our resilience by hampering our immunity. When you're under constant stress, for example, your brain and body are flooded with the stress hormone cortisol, which has a negative impact on white blood cells—the first line of defense in our immune systems.

In contrast, when we are more balanced, with our goals and behavior aligned with our deep self, we are primed for success. We are less likely to be sidetracked by anxiety or negativity, and stress hormone levels are likely to be lower so we will be healthier with stronger immunity, avoiding frequent minor illnesses all the way up to major health scares. Mood-regulating hormones and feel-good endorphins allow The Source to flow more freely.

I cannot emphasize enough that I see examples of this on a daily basis in the people that I coach. Many driven and successful people believe they thrive on stress, and that a constant state of heightened adrenaline and cortisol is a given for anyone who wants to do well in life. They are highly likely to ignore their body's signals that it is struggling to cope, whether they're experiencing palpitations, a feeling of overwhelm, digestive problems or low mood. This may go on for years. My first job is to tell them that they can't afford to ignore these symptoms, and they need to get to the root of them: the disconnect between what their heart and gut is telling them, and the rigid ideas that are propelling them onward regardless, with no regard for the physical and emotional cost. My first job is to convince them that they need to stop and take stock, listen to body and mind together and get back in touch with what it is they really want from life.

To me, the intention point as a metaphor that describes integration between our brain and body is the hallmark of

The Source operating at its full power, and it's something we should all strive for. Once we are able to integrate our mind and body more fully (and we'll explore this in more detail in Part 3) our motivation and energy will align in a powerful way.

Set your intention

I'd like you to now set your intention: the overarching goal that underpins everything else you want to achieve for the future. It should feel bold; a big-picture aspiration that reflects the area of your life that you want to change. Write this on the first page of your journal.

The idea of manifesting this change should be exciting and motivating. When you close your eyes, and imagine it becoming real, you should be able to see the picture in your mind and feel it in your gut; it should make your heart swell with desire. It could be along the lines of the following:

- Develop the confidence to build a flourishing business and find a great life partner.

- Let courage and vision govern my decisions from now on and let go of fear.

- Turn around a difficult relationship/family situation and master emotional regulation.

- Find happiness in life through greater health and life purpose.

- Adopt an attitude of compassion towards myself, and tune out my critical inner voice to create the life I dream of and which I deserve.

Don't feel constrained by your goal—aim high. As you work through this book you will learn to harness The Source to help you achieve anything you might want. We'll begin to explore how to envisage your innermost wishes in the next chapter (page 51). If later you need to adjust or clarify your intention as you progress through the book and create your action board, that is fine. You could even start collecting images for your action board as you read through the book and begin to gather ideas about what your ideal life might look like.

Now, let's take each of the six principles that underpin the law of attraction in turn and see what modern neuroscience tells us about them.

Principle 1: Abundance

The idea of tapping into the "abundant universe" is at the core of the law of attraction, and it's the first principle to take on board if you want to harness The Source. Because the "a" word has been bandied about by the most unscientific of inspirational gurus, you may have been tempted to dismiss it as self-help candyfloss. But if we take a serious look at it, it's possible to take a common-sense view that is informed by science.

The internal battle for abundance

In most people's minds, there is a battle going on between two perspectives: abundance and lack. These are like two roads that we can choose to walk down, each giving us vastly different experiences of life.

Abundance correlates with positive thinking and generosity, with the central belief that there is enough out there for everyone, and that by carving our niche and claiming our success we will add to the realm of possibility. Abundance feeds our self-esteem and confidence, helps us stay resilient during the tough times and is infectious and generative, creating a flourishing environment and community around us. Like attracts like and if you look around you, you will find positive, confident people who are friends, partners or business partners with similar mindsets.

On the other hand, when we think from a perspective of lack, our primary motivation is fear. We think in negatives, are highly attuned to what we don't have and what won't work, as well as the deficiencies of our self and our situation. We think in black and white and shrink from obstacles and limitations, retreating to a conservative, protective comfort zone, avoiding risk and resisting change. "Better the devil you know . . ." we say, or "Out of the frying pan, into the fire." We often have no more actual evidence that bad things will happen if we take a risk than we do that good things will happen if we act abundantly.

Think about your own life. Have you ever stuck with something—a job you were unhappy in, a relationship that was dysfunctional, a friendship you had outgrown—because you feared uncertainty and change? Do you worry about failure if you embark on something new? This is often the case when we've had our fingers burnt before—a single friend of mine who desperately wants to meet a partner but stopped dating recently after a run of bad experiences springs to mind as someone firmly stuck in a lack mindset. These negative pathways in the brain are strengthened as we continue to respond to life as if the worst is going to happen.

Fear is a powerful emotion, and one that occupies a primal part of our brain. In this state, the parts of our brains which combine emotion and memories become overactive with red alerts, dredging up bad memories and past failures as part of a safety mechanism to protect us from danger. This creates a feedback loop that triggers a response that is tailor-made to help us run away from risk.

Interestingly, losses have twice as powerful an effect on our brains than the equivalent gain, so we are more likely to go out of our way to avoid a potential loss than we are to try to gain a reward.[1] Blame cultures in businesses rely on this behavioral bias because people are too fearful to question poor decision-making and challenge the status quo. Remember that last time you asked for a raise and your boss hated you? Or that guy you really liked who ghosted you after three dates? "Put yourself out there again and there is a very real danger these things could happen again," says our brain, doing what it thinks is best for us. However, living with a lack mindset gets in the way of positive change, keeping us stuck and stagnant. It makes us cling to what we have, because we are hyper-aware of what we don't have. We fear losing anything and become intensely risk-averse, and a brain that is overly attuned to threat can't facilitate flexible and abundant thinking, or engage in full brain–body decision-making.

Importantly, our mindset will adapt depending on the context. We all swing from one to the other across different areas of our lives depending on stressors that may trigger a particular view of a situation. For example, most people's appetite for risk reduces significantly when they are under chronic stress. When working on a big and complex project with long hours before an important deadline, we are unlikely to take advantage of a great opportunity to buy a new home or think this is a good time to get serious about dating. This is a natural and, to a certain extent, rational response, but when you consider that many of us live in a state of near-constant stress, it's easy to see how a lack mindset can take over, leaving us stuck in a rut and unable to progress to the next level.

A lack mindset can also become ingrained in a particular area of our life, irrespective of immediate stressors. Think about your own life and ask yourself where you most practice an abundant or lack mindset: in relationships, work, friendships, or generally trying new things. Think about how this is affecting your life now and your future dreams.

So, how can we change the way we think to allow for abundant living? Abundant thinking relies on a willingness to change our patterns of thinking and make space for the new; to let go of past beliefs and assumptions and take on board new evidence and ideas. Neuroscientists have had to walk the talk on this: the advances in research have meant that things we thought to be true are no longer supported by evidence. The list of concepts we've recently had to reassess includes the brain being "set" by adulthood, the fake news of left- versus right-brain thinking/lateralization, the nature of differences in male and female brains, and the biological basis of sexuality, to name a few.

By definition, if you stand for science, you stand for being comfortable with failure and moving forward with an appetite for new learning and continuous improvement. In life, as in science, progress comes about more easily if we are willing to let go of past beliefs and embrace change. Striving for transformation in a personal sense requires unflinching honesty about our own thinking, and a willingness to change our mind.

CLAIRE IN LOVE VERSUS CLAIRE AT WORK

I have an old friend, Claire, who has an admirably abundant approach to relationships and friendships. She recently left a long-term relationship that had become dysfunctional, and was able to embark on dating with an optimistic lack of cynicism and a sense of fun that is remarkable. She has a varied collection of good friends and is great both at meeting new people and nurturing her existing friendships.

However, her work life tells another story. She has been stuck, miserably, at the same company for years. Passed over for promotion and regularly dumped on, with senior executives giving her their most thankless tasks, she has complained continuously about this job to anyone who will listen for the past four years, but seems unable to leave. Why? She lived a childhood with two freelance parents in and out of employment that has given her an enduring fear of financial instability. Coupled with a traumatic experience of being made redundant in her first big job, which has primed her to expect the worst, she clings on to the "safety" of the familiar in her work life. When it comes to her career, her lack mindset is definitely in charge.

Helping Claire to raise this long-standing negative belief pattern and how it was influencing her into her conscious brain has allowed her to make a proactive choice about how to face her career insecurity fears. We played around with the idea of applying the same joie de vivre she has towards people and her relationships to her work life. After a period of experimentation, she found a way to apply this that was authentic in her heart and mind. She realized that her skills at building good relationships and her enjoyment of networking would mean she would rarely go without work unless she wanted to.

Armed with this confidence she was able to bring an abundant mindset to her career, and she began to make choices based on what she wanted to do rather than on protecting the status quo at all costs.

We can all be a little bit like Claire, although our "lack" spot is just as likely to be our romantic relationships, well-being or social life as our professional life.

Ask yourself whether you have any deeply held beliefs that could be preventing you from making the changes you say you want to in any area of your life. For example, perhaps you are overloaded at work and say that you want to delegate tasks, but then find reasons not to do it because deep down you enjoy having the power and control of being the only person with the knowledge, and you fear that sharing your workload might mean that someone else might do your job better. It feels like too much of a risk, but is it? These are the kinds of excuses we make for fear of failure, and they are exactly the actions that embracing The Source is designed to identify and avoid. You may not be consciously aware of them, but asking yourself what lack mindsets are shaping your life is the first step on the path to uncovering the limiting subconscious beliefs that could exist in your life.

Choosing abundance

Living with an abundant approach, in contrast, is underpinned by the belief that there is always potential for improvement. To the abundant mind, challenge, learning and difficulty are innately rewarding; ends in themselves, as well as being the key to improvement and growth. Intelligence, creativity and skill—whether in art, problem-solving or relationships—can be practiced and improved on. Small failures are reframed as opportunities and part of the journey: a knock-back at work teaches you to improve a vital skill, or a romance that fails helps you to understand more clearly what you want in a partner.

This is where the power of unlocking The Source lies. Choosing an abundant mindset is a commitment to fully engage with life: to be active rather than passive and to firmly turn off our autopilot. Getting back on the dating scene, moving or going traveling may be changes that you willingly take on. Heartbreak, budgeting issues or fertility problems may be issues that you are faced with when you least expect them, but you can future-proof yourself if you can manage your fear of change. The most common effect of facing difficulties is a strong desire to stay within our comfort zone, precisely when we need to broaden our options and patterns of behavior.

What is going on in our brains when we live with an abundant mindset? Teachers and child psychologists have long understood that praising positive behavior is a better way to improve discipline, encourage hard work and create positive habits than punishing bad behavior. This is true in work and relationships too, but it can feel like surprisingly hard work when our brain didn't get the memo. To varying degrees, we all have the negative habit of picking on what isn't going well due to the risk aversion gearing of our brain being stronger than the reward gearing. A major aspect of abundance is positive thinking. This means focusing on positives rather than dwelling on negatives, overwriting negative thoughts with positive affirmations, cultivating trust and generosity towards others, and believing that life is good and conducive to enabling you to thrive.

Cultivating abundant thinking is something you need to commit to and consciously work at. As we'll discuss in Chapter 4, changing habituated patterns of thinking (both conscious and unconscious) requires effort and repetitive practice. Where lack thinking is deeply entrenched, there will be multitudes of neurons and neural pathways where a cascade of worst-case scenario "if . . . then . . ." thinking has become second nature.

Reframing failure

One of the simplest ways to begin thinking more abundantly is to change the way you consider failure. With a "lack" mentality, failures are taken up by a "told you so" inner critic and used to bolster the belief that there is little point in persisting when it comes to ambitious goals. Abundant thinkers, on the other hand, regard failure as an essential element of success.

Some of the world's greatest innovations came about in unlikely, experimental ways. Everything from Teflon and plastic to the microwave were discovered in failed attempts to create something else entirely. Charles Eames created his iconic chair as a side project—a spin-off following the honing of his innovative technique for molding plywood for leg splints. In 2003 Jamie Link, a female graduate student at UCLA, accidentally discovered what we now know as "smart dust" when the silicon chip she had been working on was destroyed. In the remains of what was left, she discovered that the individual parts of that chip still functioned as sensors. Today these are used in everything from medical technology to large-scale eco-detection. Viagra—one of the world's biggest-selling drugs—was originally developed for the treatment of high blood pressure and chest pain due to heart disease. These are all major examples of discovery through experimentation and "failure."

Reassessing our own "failures" and rebranding them as "not yets" is a good way to start rewriting our own story: the internal narrative of our past struggles. When we decide to switch to abundant thinking, there is always a positive spin. Such is the stuff of success. It means we're able to maintain the resilience to stick with our goals, rather than walking away at the first hurdle.

"Because you're worth it" has become one of the catchiest advertising phrases of recent years, for good reason. Too few of us feel, deep down, that we truly deserve, and can create, the lives we dream of, but we all wish we had the power and freedom to do so. By choosing to look at the world through a filter of abundance, and turning your back on lack thinking, you're well on your way to replacing self-doubt with the self-belief and new reality that you crave.

Principle 2: Manifestation

Despite the fact that we have all experienced this serendipitous phenomenon to some degree—you think about a big group vacation and then a friend books a large house in France and emails to invite you, or you become interested in something related to your work tangentially and then a major project in that exact area lands in your lap—it seems incredible to believe that "merely" directing our energy towards our deepest desires and focusing our attention on this can help us "manifest" our ideal life. These examples are rare occurrences and I certainly don't advocate making a passive wish and expecting the rewards to come flooding in. But, a strong intention coupled with sufficient action can make these things happen. You can ask a group of friends to pool together for a vacation, and you can let your network know what kind of work you are looking for. Often these things don't manifest because we don't have the confidence to ask.

Look at those around you for examples of when this has happened. Don't just focus on the obvious success stories of friends or family who have created successful businesses or climbed mountains, but also look at those who perhaps have made huge health changes to their lives or who found the perfect house for their needs through talking to someone they met randomly. There are some interesting high-profile examples of this working, too: from the actor Jim Carrey writing himself a fake check for $10 million, dated 1994, to go on to land *Dumb and Dumber* that year with a fee of exactly that amount, to Oprah Winfrey's life-changing vision boards.

To commit to actively trying to "manifest" our dream life may seem crazy. We fear it won't work and the effort will have been in vain, or that we will feel humiliated if we share our big ideas with someone and don't get a positive response. So we just sit back, do nothing and wait to see if it might happen without believing it could.

Too often, our deepest desires and the intentions we choose are at odds with each other. We touched on this on page 20 when we talked about setting your intention. Consider the examples of this that apply to you. Perhaps you focused on aiming for a promotion and a pay raise to achieve stability when your real dream was to retrain; or you returned to a relationship you had been miserable in because you felt you ought to be able to make it work. Think about your life and the last time that you truly "worked towards" something that was your heart's desire. What happened?

The science of manifestation

If our desires and intention are truly aligned, we can begin to "manifest" the life we want by engaging all our senses in the imagining and visualization of it—saying it; hearing it; visualizing what it looks, feels, smells and tastes like. In this way, our dreams begin to feel tangible to our brain.

In finding this focus and fully identifying it in our mind, there are two physiological processes going on in the brain simultaneously that explain this powerful cocktail and why manifestation has real effects. These processes are "selective attention" (filtering) and "value tagging." Let's explore them in more detail.

Selective attention

We are bombarded with millions of bits of information every second—mostly through our eyes and ears, but also through smell, taste and touch. Our brain must discard or fade some things into the background to enable us to focus on what is necessary to us at that time. Information is registered and stored as memories, ready to direct and influence subsequent actions and responses. Selective attention is the cognitive process in which the brain attends to a small number of sensory inputs while filtering out what it deems unnecessary distractions.

It is the part of the limbic system called the "thalamus" which manages the brain's selective filtering. During a conversation with a friend, for example, it will take in data from your visual observations (the image of the person in front of you and your observations of their movements and body language), plus the sound of their voice with its inflections and emphasis, along with any further sensory information you receive and the emotions you have in your body as you stand talking to them. Acting as a hub for our senses, the thalamus gathers all this sensory information and then, like a road traffic officer, it directs it to the appropriate part of the brain. The thalamus interacts with these other areas of the brain to stay informed about what is deemed a priority and what can be faded out. And the level of selection happening is rather astonishing . . .

Have you seen the video of the 1998 "door" experiment made famous by the psychologists Daniel Levin and Daniel Simons?[2] In the experiment, a researcher approached a pedestrian to ask for directions holding a map. As the pedestrian takes the map and is showing the researcher where to go, two workmen carrying a door pass between the researcher and the pedestrian, and a second researcher is substituted for the first. The pedestrian is left talking to another person entirely. Remarkably, 50 percent of pedestrians in this experiment didn't notice that the person they were talking to had changed once the door had passed. They were focusing on the map and directions and their brains failed to register that the asker looked and sounded completely different. Their thalamus had decided the appearance of the researcher was insignificant, fading out all sensory information that related to this. Levin and Simons have done a series of other similar experiments (you may have seen the one with a person in a gorilla suit in the basketball game).

This selective attention is happening every second. In fact, we choose to utilize it ourselves sometimes when we close our eyes to try to remember something specific, or put our hands over our ears if we're trying to concentrate hard. Understanding and accepting that we are all blocking huge amounts of information—and of course very much choosing to focus on other information—is crucial to the power of manifestation. It is a powerful reason to take charge of what you pay attention to and what you don't—you can't manifest what you don't consciously notice.

The brain's capacity to focus is not to be underestimated. Once we appreciate that our brains are selecting information to influence our actions (and "deselecting" others) then we start to appreciate the level of unseen happenings that just might be really important to our intentions, if only our conscious brain was in the know. Are you confident your brain is choosing well when it comes to what you should pay attention to and what you should ignore?

As we've already learned, the brain constantly returns to its default, which is simply to keep us safe so that we survive (see page 27). A great deal of brain energy is focused on working out who is friend or foe as this was critical to our survival when we lived in tribal times. Conversely, in the modern world, we need to actively direct our brain to move away from prioritizing these unconscious biases, and to being more open, flexible and courageous about pushing ourselves towards our goals and choices that feel "new" and "dangerous." Focusing on what we do want rather than what we need to avoid in order to survive will mean we are more likely to manifest it (in the same way that if you're mountain biking, you should never look at the potholes and boulders you don't want to ride over, but instead focus on the path through them).

The limbic system also has the job of deciding what we should retain as conscious thoughts and memories, which explains why it is also so important to raise our aspirations and future plans from non-conscious and vaguely defined to fully conscious. For example, imagine that you purposefully create a list of core attributes that you would like in a partner—qualities that resonate with you and come from your own personal experience and desires—and then you take quality time to look at the list regularly and fully explore what these characteristics mean for you. You are priming your brain to be on the lookout for and consciously sound an alarm at anyone related to your stated desires. Where previously you may have unconsciously filtered out opportunities to meet for coffee or talk to someone who seemed interesting at the bus stop because you had given up on meeting "Mr. or Miss Right," you will be more likely to notice a lingering look, an inviting smile or to actually contact someone who gave you their business card. This is why focusing your attention on your desire is part of manifesting your dreams.

Value tagging

As part of selective attention, value tagging is the importance your brain assigns to every piece of information it is exposed to—people, places, smells, memories . . . you name it. It is an unconscious activity that precedes every action in response to a stimulus and therefore directs your ensuing response.

One person will note an old red Mini parked on their street and recall fond memories of their (similar-looking) first car, and smile at the thought. Their subconscious value-tagging system is tapping into a very old memory that may have been long forgotten on the surface but still triggers a warm feeling deep down by retracing the association that was laid down in adolescence. They may take particular notice of the person they see parking the car, and start a conversation they otherwise wouldn't. Somebody else, who doesn't have a "red Mini tag" in their brain (thalamus and limbic system) may not notice the car at all, even if it is parked outside their house for a few days.

There are logical and emotional elements to value tagging. The logical element is literally about tagging all the data our brains are bombarded with in order of value to us and our survival. The emotional element has more to do with assigning value to our levels of "social safety," which is our sense of belonging in our community, family, etc., and the meaning and purpose that build up our personal and work identities.

Because of this process, it's easy to assign a disproportionate value to things we care about or a negative value (aversion) to things we fear or where we feel uncertain. For example, if someone has been through a painful break-up or simply been single for a long time and their biological clock has been ticking, then their value-tagging system may, paradoxically, become biased against looking for a companion or having children (aversion). This is where the little voice in the head starts saying they've lived alone too long to share their space with anyone or that their career or social life is too important. Henceforth, they won't be alert to the opportunity of a likely candidate for a relationship, but would be primed to see a promotion possibility in the workplace. Can you see how the brain is steering them down a path not of their choosing, and further from their dreams?

Self-esteem issues resulting from a childhood where we were criticized at home or school or labeled as a non-achiever may mean we sabotage career opportunities because, at a deep level, we fear that we are not deserving of them. Similarly, if we start a healthy eating plan but believe that we won't be able to keep it up, we can find ourselves easily giving in to temptation and making bad choices. This is because strongly emotional experiences that have shaped our brain pathways can derail our value-tagging system, skewing it towards what we think keeps us safe even if this is not conducive to thriving in our current life. Our selective filtering will prioritize avoiding shame or criticism over potential career success or romantic fulfillment.

Quite simply, when you do allow your brain to be conscious of and focus on what you want in life, the raised awareness that results will work in your favor to automatically bring opportunities into your life. It's not magic—it's just that you are able to see the possibilities to move forward with your dreams in a way that your brain was hiding from you previously.

Principle 3: Magnetic Desire

Positive desire coupled with emotional intensity attracts real life events to match. In 1954 Roger Bannister ran the first sub-four-minute mile, even though experts thought it was not humanly possible and, indeed, was dangerous. Bannister himself was convinced it was feasible though, and once he achieved this feat, several other athletes matched it in quick succession soon afterwards (his fiercest rival John Landy less than two months later). What had changed? There wasn't suddenly better equipment or facilities to allow it to happen, but simply once it became a real possibility in people's minds, they were able to do it again. We know that merely registering that something is possible in the brain can change what happens in the body or the outside world.

Magnetic desire is a useful idea in a metaphorical sense, but it's important not to be overly literal in your understanding of it. The literature on optimism and a positive appetite for change and risk shows that an individual's mindset and determination to achieve their goals dictates what happens to them: whether they take risks, make positive changes and how they interact with others. A study at University College London (UCL) found that after a heart attack, optimists were far more likely to take positive lifestyle changes on board—they were more inclined to give up smoking and up their intake of fruit and vegetables and make changes to their lifestyle than the pessimists.[3] As a result, the optimists' risk of suffering a second heart attack or serious disease dropped significantly. Pessimists were twice as likely to experience a second severe heart attack in the four years after their first. Merely seeing the opportunity to change the future and being positive about the potential outcome had a huge impact on the optimists' future.

It's a fact that both expected and unexpected things are going to happen to us and it's how we respond to them that matters. Positive desire is the mentality that we can make good things happen, and it's the emotional intensity of that desire that drives it towards a tangible outcome. Intense emotion gives us renewed energy and confidence to carry out new actions that turn the positive desire into reality, rather than stagnating at daydreaming or hoping in vain.

Making it happen

My own journey to magnetic desire has been a work in progress, with the toughest times absolutely becoming the biggest turning points. I was in my mid-thirties when I made my career leap—probably the biggest change of my life. I turned my back on life as a National Health Service doctor, where I was part of a huge organization in a safe and stable job with no room for errors or uncertainty (this was literally life and death) on a modest but regular paycheck. I did this with no new job to go to and very little financial back-up, deciding to retrain completely and start over. But for years it didn't occur to me that I could ever be anything other than a doctor.

Beneath the surface however, there were changes afoot. My neural pathways had been growing and changing over the two years it took from first thinking about it to making the leap. All the emotional intensity of the personal change I was going through came to a head as I acknowledged my suspicion that both in terms of intellectual challenge and life experience, psychiatry couldn't offer me the mental stimulation and sense of purpose I craved. I started being more vocal and energetic about setting up my own business. To build my confidence around my positive desire to change career I read a book called

Working Identity by Herminia Ibarra filled with stories of successful career changes and I made a list of 100 things I could do instead of medicine. Even though only one of them was viable, it was enough to kick-start a process of new action. Over a number of months, I turned my vision of creating a successful and meaningful career outside of medicine into a possibility and, slowly, that possibility became a reality.

The more certain I felt inside that I needed to find something else to do, the more my confidence grew on the outside. I sought the advice of a few mentors, confiding the seed of an ambition I had to become a well-being coach, but I said it was just a dream. I knew nothing about running a business! Nonetheless I woke up one day and knew the time had come; my neural pathways had reached the tipping point. I had created the perfect storm of positive desire and emotional intensity to make a real change in my life. Luckily, I had saved up a few thousand pounds and I made the arrangements I had to.

I quit my job, signed up for a coaching course and moved back to London on my own after two years living abroad with my soon-to-be ex-husband. This was in 2007. While my marriage disintegrated I had to build a business from scratch. What I was not mentally prepared for was facing the abyss of my money running out and having a handful of clients paying friends and family rates, as the only way I could get any work at first was through recommendations from friends. The concept of networking was alien to me, but I took to it like a duck to water because I had such a strong desire to make my new business work. By dedicating energy to it consistently I slowly built up a few corporate clients in 2008. I set a vision that by 2011 I would be working as a successful coach, with a good range of clients and varied projects ahead of me, including some speaking and writing. I determined too that I'd be earning more than

I ever had as a doctor because this was a tangible measure of success and would help confirm that I'd made the right decision. This was my magnetic desire and it attracted the real-life events to match.

Along the way I had to live with my parents and my best friend's parents, and when I finally started renting a studio flat, I sometimes had to accept money from my ex-husband to pay the rent, which filled me with shame and fear for my precarious future. People urged me to work one weekend as a locum doctor to replenish my bank balance, but I remained firm that any return to medicine would feel like I had failed and crush my confidence—a precious asset that I clung on to and a radically new action for me backed up by a strong drive to succeed. When I spoke to my friend Jo, I shared my worries around not having any clients and totally running out of money. She had always worked freelance in TV and she told me "work always turns up." I chose to believe this. It did. Both the beliefs and the outcomes topped up my drive and determination—this is magnetic desire in action. It is self-perpetuating.

I learned to remain flexible and open to opportunities that came my way. I slowly increased my prices. I started working all over the world. I moved from only coaching to also accepting paid and unpaid speaking engagements. I incorporated new technologies into my coaching work and designed a signature mental resilience program for teams. I went from being solo to having one, two and then more people on my team. I set a vision of earning as much from speaking as I did from coaching. Now it's twice as much. When I set out on my new path and turned away from medicine, I imagined a future where I had variety and balance: some reading around neuroscience topics of interest to me, some writing, some coaching and a beautiful home where I could work without impinging on my personal life. I'm pleased to say that this is now a reality beyond what I could have hoped for, and the distinctiveness of my new life really fulfills me and my personal and emotional needs. Once you feel the power of magnetic desire, it multiplies with each new iteration and nothing that you have dreamt of feels out of reach.

Principle 4: Patience

Despite a pure intent and focus on what it is we want to achieve, sometimes we either give up too soon or get anxious and desperate for the process to work.

This principle is about enjoying and, most importantly, trusting the process and allowing things to unfold naturally in their own good time, rather than being obsessed with goals and achieving them. Strengthening The Source through the practices of visualization and action boards involves skills that will improve over time as you build and strengthen the pathways in your brain. Before we can get to the practical exercises to do this, though, it's really important to fully understand the underlying principles. Things like changing your attitude and becoming more confident, trusting and open to trying new things *may* take even more time. As you build these new neural pathways there seems to be a point where it feels like nothing changes for a while, then suddenly things fall into place and everything becomes more effortless. A friend of mine experienced this recently with her new business where months of cold-calling and relationship building finally paid off just as she was ready to throw in the towel. A lot of effort and resource goes into connecting neurons and building new pathways in the brain. Progress seems slow and then there is a watershed moment. After that the process and the impact of it gathers pace. This is because there is a critical mass effect for new healthy behaviors, but once you get there it feels effortless.

Similarly, any on-boarding of a new skill takes targeted effort and repetition. The feeling that you've finally "got it" and turned a new habit into second nature is a sign your brain pathways have reached that critical mass.

Principle 5: Harmony

The principle of "harmony" teaches us that in order to fully tap into and access the insights, power and gifts that life has to offer through The Source, we need a balance between our mind and body, and the knowledge that they are connected. This is a skill that is repressed by the "living in our head" lifestyle of the modern world where our bodies are disregarded as vehicles that merely carry us from meeting to meeting or from relationship to relationship. We need to be able to be more fully present in our body and brain together in the moment to find the balance and strength to make the best choices and to help our emotional regulation. This is why mindfulness and presence are an important part of The Source (see pages 216–23).

Understanding and aligning messages from our logical brain, emotional brain and gut (the idea of mind, body and spirit being on the same page rather than in conflict) are the foundations to living in harmony with ourselves and thriving in a world that is constantly evolving. Only then can we trust our feelings and have the confidence to sense what is right and best for us and within our communities. These "messages" can range from getting goose bumps when we feel unease to a sense of peace when a situation aligns with our deepest desires and core values.

Noting in your journal what happens when you follow your intuition/body instead of doing what is expected of you, or what everyone else is doing, can be an eye-opening exercise. Even minor deviations from your own needs (such as acquiescing to your partner's choice of vacation destination or going to a work event because you feel you ought to) have hidden costs. If there are lots of these compromises, it can undermine your deepest needs and aims, stirring up resentment and anger that lead to stress, elevating your levels of stress hormones which prime you for threat and push you into survival mode, making it harder for you to think from a perspective of abundance, and to fully form and focus on your true intention.

There are two major systems we can hone that will help us to gain access to our deepest intrinsic wisdom and personal power: listening to messages from our body ("interoception") and paying attention to what our "gut instinct" is telling us ("intuition"). We look at what is happening in our brain and body during these in Chapters 7 and 8. Later in Part 4, we will learn how to use the practical tools of journaling, gratitude lists and mindful living as the conduits to achieving this power.

Principle 6: Universal Connection

This is the idea that we are all connected, to each other and to the universe. It is this principle that underpins an abundant attitude to life.

As social creatures, we have a great need for belonging. The desire to foster alliances and act in a compassionate and collaborative way with other people and the world is a powerful motivator in neuroscientific terms because it activates the brain's empathy pathways. Attachment emotions such as love and trust trigger the release of the neurochemicals oxytocin and dopamine, which contribute to feelings of bonding and pleasure as part of the brain's reward system. Countless studies show that having a strong sense of meaning and purpose correlates with life satisfaction.[4]

Living in a way that is beneficial to us and is harmonious with others and the universe is better than directing our energy "against" other people or circumstances. By striving to do this, we make decisions that boost not only ourselves, but also those we are responsible for. In a broader sense, it reminds us of the responsibility we all have to the vulnerable and less privileged within our world, which is embedded in the moral circuitry of the brain.

This principle is as much about the way the world impacts on you as it is about your impact on the world. The quote that comes to mind in terms of how you apply this practically in your life is from Mahatma Gandhi: "Be the change that you wish to see in the world."

Given the latest research on neuroplasticity—the incredible ability of the brain to change itself well into adulthood—we should focus less on being the victims of outside influences and more on making proactive changes in our lives that will have an inspiring and motivating effect on those around us. Getting therapy to improve personal relationships or applying for a promotion rather than holding back or waiting to be asked are examples. This extends outwards to include a beneficial effect on society, conservation, the climate and anything else connected to our work and relationships.

Think about what it is that fires you up—what you feel passionately about—and pick one thing that you could do that would contribute to improving this. It could be as simple as recycling in your own home or volunteering a few hours a week for a local charity. Other easy ways to tap into universal connection include raising awareness on social media of injustices somewhere in the world, setting up a regular donation to an organization you care about, offering to support an elderly neighbor or training for a sponsored sporting event.

Consider your "tribe"

Our need for social connections is primal—the people around us form our tribe—and The Source relies on them to thrive. It is crucial to remember that the quality of these connections has a huge influence on our thinking, mood and behavior. For small children, this is limited to immediate family, but as we grow up, it expands outwards. Adults have the freedom to redefine this tribe, curating the connections we want—growing, pruning or allowing them to wither from disuse; exactly the same thing that happens to connections in the brain.

The word used by psychologists and sociologists to describe the impact of our social connections on us is "contagion," and there's a growing body of research on the topic. Studies show that we are influenced by those closest to us in a wide variety of ways: from lifestyle habits (both healthy and unhealthy) to emotions and even our finances. For example, a close friend getting divorced has been found to increase your own risk of divorce significantly.[5] Similarly, research shows that if a friend becomes obese, your own risk of obesity increases by 57 percent the following year.[6]

The contagion of stress has been the subject of recent research by a team at Hotchkiss Brain Institute at the University of Calgary.[7] In their study, the mating partners of stressed mice showed similar changes in the neurons that control the brain's response to stress. Even in humans, we still see evidence of women who live together or work closely together synchronizing their menstrual periods in two to three months. Through a similar mechanism we also influence each other's stress hormone levels by suppressing our own stress.

Try the following exercise to examine the impact on you of the people you spend most time with.

The people tree

1. In your journal, draw a tree with five branches and, on each branch, write the name of one of the five closest people to you. These could be a mixture of friends and family and colleagues, the people who feel most significant in your life at the present moment.

2. Along each branch, write five words that best describe that person. These can be positive or negative, and should sum up that person and what they mean to you.

3. It is often said that we are a combination of the five people we spend the most time with, so take a look at these words and see how much of them you recognize in yourself. Put an asterisk by words that relate to any strengths of your own you recognize and draw an X by the negative traits you share.

4. Think about how you can ameliorate the Xs. We are often most judgmental about others for things that deep down we fear in ourselves.

Once your tree is complete, look at it and at the 25 words you have chosen—these traits are influencing you constantly. What is the impact of these people on your mindset? Are your interactions with them likely to be fueling or draining The Source?

If your tree is filled with negativity, you'll need to take action to change it. Could you see these characters less, or change the way you interact with them to minimize the negative effect this has on The Source for you?

Ask yourself who brings out the *best you*, and who needs to go? Jot down notes in your journal as to three steps you should take to move these relationships forward into ones that support your desire to change your future. Pick one person you will spend more time with and learn from; one that you will continue your relationship with exactly as is, in a mutually rewarding fashion; and one that you intend to prune away proactively or allow to dissolve naturally by disconnecting mentally.

This exercise and the small ways to pay-it-forward will make you feel more connected with the positive energy of other people, which is a great way to bolster your own. This is the energy that the law of attraction thrives on, and we must be its generators as well as its conduits.

The six laws of attraction, reinterpreted in the light of cutting-edge science, are tools that will help you on the way to realizing the full power of The Source. Manifestation and magnetic desire are useful for raising your awareness of what you want and focusing your attention on it, guiding your actions to make it happen. Patience and harmony will help ensure you stick with your goals and that they align with your deepest self. Finally, becoming aware of abundance and universal connection encourages you to think about your goals in the context of other people and the wider world; to consider your place in it and provide you with a powerful sense of purpose that will guide The Source, making you more resilient, compassionate and integrated in your thinking. This shift leads to an exponential increase in the consciousness of your own power.

Chapter 2

Visualize It

"If you don't know where you are going, every road will
lead to nowhere."

Henry Kissinger

Wwithout fail, in the run-up to a big competition, skier Lindsey Vonn, a multi-Olympic gold medallist, visualizes herself skiing the course:

I always visualize the run before I do it. By the time I get to the start gate, I've run that race 100 times already in my head, picturing how I'll take every turn . . . Once I visualize a course, I never forget it. So I get on the right line and go through exactly the run that I want to have.

Visualization is a technique a lot of athletes employ. Everyone from Muhammad Ali to Tiger Woods has spoken about visualization as a big part of their mental preparation for competition. Outside sport, countless celebrities also credit their success with visualization. Examples include Arnold Schwarzenegger and Katy Perry, who was once photographed alongside the vision board she put together when she was nine—all of which, including winning a Grammy, had come true.

The language of self-belief and achievement is rich with visual metaphors. We "dream" of doing something great or we see something happening in our "mind's eye": this is the language we use more when we are in touch with all our senses and comfortable with daydreaming and mind-wandering rather than focusing on rational thought and concrete examples only.

Visualization works because there is surprisingly little difference to the brain between experiencing an event directly in the outside world and a strongly imagined vision (plus sometimes imagined action) of the same event.

The Power of Visualization

Let's start with a simple example: imagine tapping your left foot on the floor. By doing this you have stimulated the same part of your brain that is active when you actually carry out that action. There are even brain scans that show people in a coma (who cannot move or respond) being asked to imagine walking into their living room and activating parts of the brain that are related to walking as well as imagining.

Incredibly, simply imagining something can also deliver physical as well as mental benefits of the imagined action: something can begin to feel and even *be* real purely by imaginative means. Studies show that people who imagine themselves flexing a muscle achieve actual physical strength gains and activate brain pathways in the cortex that are related to the movement.[1] Similarly, when exercise psychologist Guang Yue asked a test group to do imaginary workouts at the Cleveland Clinic Foundation in Ohio, his team found increases in muscle mass despite not doing any activity.[2] Incredible! Thirty young, healthy volunteers participated in the study. The first group of eight was trained to perform "mental contractions" of the little finger; the second group of eight performed mental contractions of the elbow; and the third group of eight was not trained but participated in all measurements and served as a control group. Finally, six volunteers performed physical finger training. Training lasted for 12 weeks (15 minutes per day, 5 days per week). At the end of training, the first group had increased their finger strength by 35 percent and the second group improved their elbow strength by 13.5 percent (despite taking no actual physical activity!). In contrast, the physical training group increased their finger strength by 53 percent. The control group showed no significant changes in strength for either finger or elbow tasks. Despite the fact that the group who did physical exercise saw greater benefit, the increase in the imaginary group is still mind-blowing.

This provides hard evidence for something sports psychologists have long understood: by creating a mental image of the things we want to achieve, and furthermore by matching visualizations with simulated physical sensations, we improve the brain–body connection in relation to this activity. The brain registers this at a deep level and is more likely to make a positive connection to a trigger or event related to this in real life. Similarly, hypnotherapists commonly recommend having an elastic band on your wrist that you snap whenever you complete an action you've committed to, or having three bands on the left wrist that must be moved over to the right wrist over the course of a day when you think a positive thought or visualize a positive outcome.

Bringing together a physical and a mental trigger will activate the body as well as the brain, forming double the reinforcement for the desired outcome. We are priming our brain to recognize and be skilled at something even if we have never seen or done it before. It makes sense therefore that visualizing our ideal future primes the brain to recognize aspects of it in our daily actions and interactions, see opportunities that will help us and be attracted towards them.

Visualizing in order to tackle a particular event is also hugely effective because it reverses the paradigm that any new scenario, person or place tends to be treated by the brain as a potential threat. Think of a high-stakes meeting like a job interview or a blind date. We are bound to be nervous as our brain reacts to anything we do not recognize as familiar or that takes us out of our comfort zone—staying vigilant to look out for potential danger is the brain's preferred state in the face of anything new or any changes. As we saw on pages 25–33, this state is dominated by "lack" thinking—the enemy of an abundant mindset. This is something we are all vulnerable to, particularly at times of stress, as our adrenal glands continue to release increasing levels of the stress hormone cortisol that is detrimental to our health, and more importantly in this case, it biases our decision-making to avoid risks. All these things can attack our self-esteem, and the belief systems that help us in these types of situations—therefore affecting our performance and ability to cope with new situations.

However, when we visualize a particular event or situation in advance, we essentially trick our brains into thinking that it is already familiar with the event or challenge that we are visualizing, so it becomes less mistrustful, moving towards the abundant mindset that allows us to take considered risks and seize opportunities.

When I use visualizations with people in order to prepare for a single event—such as attending a job interview, giving a speech or winning a competition—they will imagine every aspect of that event in their mind's eye. This should include what they are wearing (imagining looking down at their feet and seeing their shoes as they would on the day, the clothing on their body), who they are with (imagining the sea of faces and their reaction to the speech or interview) and where they are, playing out the scenario in full and ending it with a positive result. If people have been to the venue before they can picture themselves there; if not, they can google it or try to pop in just before the event so it is even more familiar to the brain. This is no different than looking up the route on a map or app before setting off for a long drive. We wouldn't think twice about doing this on an unfamiliar route, yet we take it for granted not to prepare in the same way for an important event.

While this form of visualization can certainly be useful for tackling a single high-stakes event, in this book we will take the power of visualization to the next level, using it to build a long-term vision for your life, and exploring the neuroscience behind it.

Visualization to Create Your Future

Visualization helps us to channel the law of attraction, and to act from a perspective of abundance and optimism. It works by raising our awareness, directing focused attention to the things we want most in life, and overcoming the brain's desire to protect itself from new or difficult situations. It is also tied to two particular capabilities of the brain: "abstraction" and "sensory integration."

Abstraction

Abstraction, or abstract thinking, is the ability of the brain to construct representations of things that are not present or definitive—to imagine possibilities, to see patterns where they were not obvious before and to join the dots. From abstract concepts, such as astrophysics, to the creative use of language, such as poetry, abstraction is the opposite of logical thinking where everything is set in stone and requires no creativity or imagination.

Abstract thinking is when we ask "What if?" when faced with a situation, and whether we can break down problems into bite-sized chunks to work out how to solve them in new ways. It allows us to identify patterns in behavior and amend our responses, to develop new ways forward and to imagine the yet-to-come, for example, our future dream vacation: where it is, what we are doing and who we are with. This may be a pipe dream at the moment but we use this combination of memory and knowledge with flexible thinking to picture it in great detail, almost as if it were real, here and now.

There are multiple and complex networks in the brain—the bookends being the "default" and "control" networks. The brain's default network is what enables us to think in an abstract way, which is the perfect counterbalance to the "can't see the forest for the trees" feeling we all get when we're mired in thinking logically and functionally about our day-to-day tasks or living under stress. Activities like aimless lounging around, daydreaming, puttering and reading for pleasure rather than purpose all activate the default network in the brain. When this network flourishes, inspiration is more likely to strike and we are better able to free-associate and harness our emotional intelligence and intuition. Perhaps that's why, so often, we return from vacation with a fresh perspective on a situation we've been feeling stumped by, or the resolve to take our lives in a bold new direction. The break helps us to envisage new possibilities: to imagine new solutions to old problems. We then have to act on this.

But our logical brain is used to being "always on." The counterbalance to the default network we just mentioned is the brain's control network—the series of pathways that governs our task focus and analytical thinking. So it makes intuitive sense that when you're trying to give this network a break and let your brain relax, free-associate and access those "blue sky" moments we all crave, you need to help switch off the control network.

Visualization is a great way of turning away from logical dominance and accessing a more abstract and flexible way of thinking.

It starts by integrating all our senses into the concept of visualization to harness the brain–body connection—and this is why I ask people to tell me what their ideal vision looks like, sounds like and even feels, smells and tastes like—and allows us to fully embrace the unknown and as yet unexplored aspects of our lives. *The Source* will help you to identify the entrenched and rigid belief patterns that dominate your thinking and help you to build in your mind alternatives that allow you to develop and move forward. The exercises throughout this book and in Part 4 are designed to raise from non-conscious to conscious these patterns in your thoughts and your behavior that may have been playing out during your life. You will then be able to challenge your own thoughts and choose new, fulfilling behaviors.

Sensory connections

The brain creates what we see as reality through the massive amounts of data it receives through all our senses from the outside world. We then use abstraction as the information triggers particular memories to create connections between these sensory triggers and past events that we recall experiencing. Smell is usually the strongest stimulator of a memory—fond or disgusting—but all the senses interact with memories in a similar way. Hence, we have the power to prime ourselves for success by using our senses to connect to memories of abundance, opportunity or fulfilling relationships.

Beginning to Visualize

When I use visualizations, I encourage people to feel as well as to visualize whatever it is they imagine. Visualizations should harness all of our senses, tapping into an imagined and "felt" experience of them. You will be able to conjure a full sensory experience by creating the feel, sound and scent as well as the visual image of your imagining.

Below is a simple visualization that acts as a powerful way to recognize the most positive and negative mental states at work within you.

Positive you, negative you

Any meaningful personal development starts with a raised awareness of yourself, and this exercise is about improving your awareness. Draw the following table twice on two pages of your journal, divided into quadrants as shown:

Physical	Mental
Emotional	Spiritual

"Physical" refers to what you feel in your body; "mental" is about what is going on in your thoughts; "emotional" is how you are feeling; and "spiritual" is about how you feel deep down, at a more fundamental level, in terms of your sense of meaning, purpose and place in the world. You will be reimagining all of these thoughts and feelings by conjuring up past memories of a time when you felt very negative, stressed or unhappy; followed by a contrasting situation when you were confident, happy and fulfilled. This should activate the same feelings that you felt at the time as you recall what happened.

1. Start by thinking of a time when you felt very stressed, your confidence was low or things were not going your

way. You might remember a time when you were at risk of redundancy or you had split up with a partner, or maybe a period when you were low or a meeting/conversation that went badly wrong.

2. Close your eyes and immerse yourself in this memory for one minute (time this on your phone so that you don't go over the minute). Take the time to relive the sights and sounds of what happened—recall details such as what you were wearing and who you were with.

3. When the minute is up, open your eyes and make notes in the four quadrants straightaway. In the physical quadrant, you might write "exhausted, stiff muscles"; in the mental box "racing thoughts" and "Why me?"; in the emotional area, you might write "sad, angry, humiliated"; and in the spiritual "lost" or "disconnected."

4. Next, recall a time when you were happy and confident, and life was good. Close your eyes and immerse yourself in this memory for a minute, timing it as before. You might remember your wedding day or a landmark birthday when you were surrounded by friends and family, and life felt full of hope. How does that manifest physically, mentally, emotionally and spiritually?

5. This time, make notes in each of the four quadrants on the second page.

6. Now compare the two sets of notes. What strikes you as being surprising or obvious, and what are the similarities and differences between the two sets? There are no right or wrong answers. Look at what is relevant to you and use this to learn how to move yourself from inertia

to confidence through an action related to one of the quadrants whenever you need this. This could be a physical mannerism, recollecting how grounded you are feeling, or very much to do with which mindset you are embracing.

Whenever I do this exercise, the physical quadrant is the most strikingly different. That's because when I am feeling down I avoid eye contact, tend not to smile and my posture becomes slumped.

What can you do to turn a bad day into a good day? To turn yourself during a difficult moment into your best self? Write down the answer to this in your journal, and make a mental note to keep your eyes peeled for images in magazines that could represent your positive self on your action board.

If I find it too difficult to change my thought patterns, regulate my emotions or lift my spirits, I know I can at least lift my chin, put my shoulders back, make good eye contact and smile at people.

Throughout this week, aim to adopt an attitude of positive encouragement towards yourself at all times. Praise activates the emotional circuitry associated with love/trust and joy/excitement which correlates to the bonding hormone oxytocin, making us feel warm towards others and ourselves. As a result, when we're in the habit of thinking in this way, we're more likely to act on the basis of abundance, and feel that no one else's success diminishes our own.

Self-criticism and negativity, on the flipside, activate the survival circuitry associated with emotions such as fear, anger, disgust, shame and sadness. This biases our perspective towards scarcity and makes it more likely we will remain frozen in the status quo rather than take any risks that may lead to further "punishment."

Think about the amount of time that your brain is effectively "reliving" previous difficult or happy occasions, and how this visualization will be affecting your mindset and decision-making. Resolve to put on your "positive hat" more often, and consider ways of making a positive visualization a regular habit. You could try spending two to three minutes a day visualizing your confident self from the exercise above and build this up to five to ten minutes or more if you enjoy it and reap benefits. You will be amazed at the effect it has on you and what this generates in terms of how you relate to others as well as how they respond to you.

Visualization is not just about creating an image of what you want but also about imagining what it would feel like if you were really in that picture. Everything from the taste in your mouth (the taste of success), the smells around you (the fresh paint of a new home, food-related smells in a certain career, your favorite perfume for special occasions), what you are hearing (applause, congratulations, music) and, hugely importantly, the physical feeling in your body of achieving this (how does happiness or confidence actually feel?).

The more we practice this, the more likely we are to start noticing when it is happening, or the things that are associated with moving us closer to that ideal. Perhaps use a specific essential oil to accompany your visualizations or mind-wandering time. The visualization exercise above, along with those on pages 228 and 229, will help you to get all your senses working for you, picking up clues and integrating your brain pathways and corresponding ways of thinking to bring your vision into reality.

PART 2

The Elastic Brain

Chapter 3

Your Amazing Brain: Genesis of The Source

t is often said that we currently know more about outer space than we do about the brain; more about the billions of light years of unknown matter around our planet than how the 1.5kg of cells inside our head works. But our brains hold untold potential in the billions of neurons firing away in there. Some of the most exciting gains in our understanding of neuroscience over the last decade or so have uncovered new information about just how much the brain can change, responding to focused effort and targeted practices.

However, the more you can understand how the brain physically works, the more you can unlock the potential of The Source, and this ability to unlock The Source is both vital and transformative, enabling us to fulfill our deepest desires; support healthy, reciprocal relationships; and plan for the future.

Old Science, New Science

Until relatively recently, conventional wisdom was that once we had finished growing physically, so, too, was our brain fully formed. We thought that no new neurons could be born in the adult central nervous system (CNS), and hence so much of our personality and potential was hardwired for life.

We've known for a long time that nerves in the arms or legs, for example, could regenerate if severed, but not the brain or spinal cord. It was understood that we would be able to carry on learning, assimilating information and memories and honing skills into adulthood, and that it is perfectly possible to change your *mind*, but changing the *brain* itself, at a deeper, physiological level, was simply not possible.

Modern neuroscience, and the advent of brain scanners, blew this theory out of the water. We now know that embryonic nerve cells *are* found in the adult CNS, mostly in the hippocampus (which makes sense as this is where we form and store new memories), and although it is still contentious as to whether these cells can be found and grow in other areas, this is looking increasingly possible. We are living in an age of enlightenment, constantly discovering new things about the brain and its remarkable processes. The brain itself, and what we thought we knew about it, turns out not to be set in stone after all.

One of the most famous stories in neuroscience is that of Phineas Gage, a Californian railway worker in the mid-nineteenth century who changed so much of what we knew about the brain after suffering a work accident. Gage was the foreman of a team of men digging out a railroad bed, using an iron rod to push explosive powder into a hole on a wall of rock when it exploded prematurely. The meter-long rod smashed through Gage's cheekbone, pierced his brain and exited through the back of his skull, landing several meters behind him and destroying most of his brain's left-frontal lobe. It's incredible in itself that he survived this.

The changes to his personality after his recovery were so immediate and profound his friends called him "no longer Gage," and the railroad company who had thought him a model foreman refused to take him back due to his appalling behavior and loss of inhibitions. Observations of Gage's behavior during the remaining 12 years of his life enabled doctors and scientists to study the brain's response following physical trauma. Gage's remarkable story was the first real case that shed light on the modern understanding of how the brain manages behavior, determines personality and governs insight. Gage suffered severe personality changes, and he was unable to make plans or inhibit his impulses. This gave rise to the idea (since proven) that the pre-frontal cortex (PFC) is vital for impulse control, and predicting and planning for the future.

We now know things that we couldn't have dreamt of even a few years ago, let alone in Gage's time. Over the last 20 years, with the advent of sophisticated brain-scanning techniques, we have started to uncover the true magnificence of the brain and its pathways. I am going to take you on a journey through your brain, from how it developed and is organized, all the way up to how it regulates everything we experience. This is what makes you YOU. This is the genesis of The Source.

The Amazing Brain

Our CNS, as it develops from birth and to the end of our lives, comprises:

- the *cerebral cortex*—the folded outer surface that is generally what we think of as the brain

- the *brainstem* which links the "brain" to the spinal cord

- the *cerebellum* at the back of the brain, which is mainly involved with coordination and movement

These parts all function together, beautifully interconnected like a completed 3-D jigsaw puzzle with 86 billion pieces, each piece representing a neuron in the human brain.

The neurons (nerve cells) are the means by which we can convey and interpret information from all over our body and our senses, and that coordinate movement, behavior, communication and thought. The neurons relay information around the brain via neural pathways passing electrical signals within the areas of the brain—from what we see, hear and feel, to our reactions to heat and cold, touch and our emotional responses.

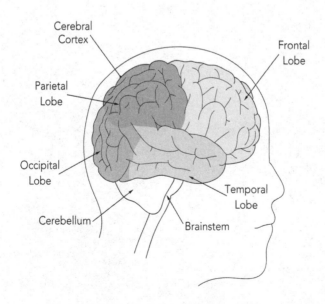

The human brain

Neurons are fascinating. They look rather like trees: with a "trunk" called the *axon*, "branches" called *dendrites* which receive information from other neurons, and "roots" called *axon terminals* which send information to other neurons in the form of an electrical message. Information moves as an electrical impulse in the "roots" of the neuron that triggers the nerve ending to emit a chemical called a *neurotransmitter*, which then bridges the gap (called the *synapse*) between it and the next neuron.

The neurotransmitter is received by another neuron's "branches," which then prompts the second neuron to transmit another electrical impulse in its nerve endings, and so the electrical impulse continues from neuron to neuron.

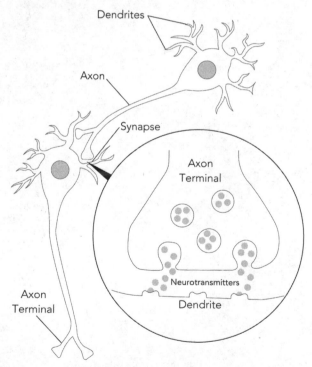

Neurons (nerve cells) in the brain

Each neuron is extremely hard-working—it can transmit 1,000 nerve signals each second and make as many as 10,000 connections with other neurons. We now know that new and existing neurons can make more and more of these connections as we grow and change our brains in response to everything that we experience well into adulthood. All our thoughts are emergent from the chemical and electrical signals that pass across these synaptic gaps between our neurons: the more connections we make, the more we are able to unlock The Source.

These signals are happening simultaneously and in different configurations all the time, even when we are asleep. In fact, while we sleep we are building pathways, making new connections and growing cells, all to process the world we experience during waking hours and then refine our responses to what happens each day as our brain changes and we become mature and more experienced. As the pathways become stronger and more established over time and with repetition, these become the habits and behavior patterns that we may or may not be consciously aware of. Neuroplasticity means we can disrupt and refine these patterns well into adulthood.

Inception of the brain

The first sign of nervous system development happens at around three weeks after fertilization inside the womb. Embryonic cells form, unfurl and then fold to form the neural tube which becomes the brain at one end and the spinal cord at the other. This is an extremely complex and delicate process guided by our DNA, and it makes us who we are when we are born.

Each tiny little human being truly is a miracle from the moment it enters our world, and what happens next, especially in the first two years, as the brain rapidly develops and we learn to walk and talk, is incredible. Babies' brains grow by approximately 1 percent each day for the first weeks after birth. Overall, they grow by about 64 percent (from about a third of adult size to over half) in the first three months.

The cerebellum, tucked away at the back of the nape of the neck, is involved with movement and critically with balance, and is the fastest growing region early on, more than doubling in the first three months of life.[1] We see the effects of the cerebellum being disturbed when people are drunk (wobbly walk and clumsy movements with a likelihood of falls just like a toddler!) as alcohol affects that part of the brain significantly.

The growth of babies' brains reflects how our brains have evolved over millennia. The folds of the cerebral cortex grow in an uneven fashion depending on what is most important in the first period of life. These folds are divided into two hemispheres—the left and the right (and we used to be convinced that this was hugely relevant to how people's brains functioned—but more on that later).

Each hemisphere of the cerebral cortex includes four defined areas or lobes:

1. The *frontal lobes* control our reasoning, planning, problem-solving and short-term memory storage, plus movement.

2. The *occipital lobes* process information from our eyes, and link it with information already stored in the brain.

3. The *temporal lobes* work with sensory information from our ears, nose and mouth, plus they are involved with memory storage.

4. The *parietal lobes* are involved with sensory information from our ears, nose and touch.

Broadly, there are visual, auditory, and even language centers and more in the brain but all functions rely on complex networks to fire simultaneously and, like a fingerprint, the maps in each of our brains for all functions will be unique and dynamic.

Association areas are parts of the cortex which are not connected to movement or sensation but are instead involved in the more complex processing of sensory information and the perceptual experience of the world. The association areas are organized as networks that are distributed throughout the brain and include temporal, parietal and occipital areas at the back of the head, and also the pre-frontal areas. The PFC is at the very front of the cortex and head and governs *logic* and *creativity*. Together, the association areas make sense of the countless cascades of information flowing around the brain and body. The PFC, which has increased in size as humans have evolved, is involved in purpose and responsiveness to the outside world, risk-taking and the ability to work towards a goal—often linked to what we call "higher-level thinking" or "executive function." When the PFC is not functioning optimally we become more distracted, forgetful, disinhibited, inattentive and emotionally erratic. We maintain the same old narrative in our head and tend to continue to repeat previous patterns of behavior even in the face of change around us. Is this sounding familiar?

Being able to consider opposing views in our mind and come up with new solutions and responses is one of the highest functions of an optimized brain. We can all learn to do this by leveraging whole-brain thinking through the association cortices—integrating our senses and using abstract thinking to see patterns that are not obvious. When The Source is functioning at its full capacity, it creates space for new connections to develop as our pathways work together, laterally and holistically. Creative thought is free to flourish, rather than being shut down by a brain that's stuck on autopilot.

The initial high-growth areas in babies' brains are linked to language development and mental reasoning (parts of the frontal and parietal lobes). The areas related to vision (the occipital lobes) that process what we see are already quite well developed at birth so the baby can recognize and bond with its parents.[2] More intangible skills—such as trust, love and resilience—are influenced by our environment and relationships. These develop over a much longer period of time—well into adolescence—as they are less critical to the initial survival of the baby.

The neural pathways connecting different parts of our brain develop and are strengthened during childhood, most rapidly in the first 12–24 months during the intensive processes of learning to walk and talk. Then, during adolescence there is a massive amount of neural "pruning" that gets rid of unused pathways and specializes the brain with sophisticated pathways for the life skills that will be needed to navigate social interaction, survival and reproduction.

The *spinal cord* connects the brain and body, with sensory information moving upwards from body to brain and movement information traveling downwards from brain to body, like a two-way information highway. The spinal cord, movement and sensation parts of the cerebral cortex relate to the crucial capability of *physicality* in the brain agility model that we will come to later in Chapter 7. This facilitates our ability to listen to messages from our body and harness the brain–body connection in both directions.

On the edge

Tucked away deeper inside the brain and about the size of our fist, is the *limbic system*, the more primal emotional and intuitive parts of our brain where our non-conscious habits and behavior patterns are stored. This is the part of the brain that is crucial for us to harness in order to fully maximize The Source. Associated with behavior, emotion, motivation and the creation of long-term memories, the main parts of the limbic system are the *amygdala, hypothalamus, thalamus* and *basal ganglia.* (The word "limbic" comes from the Latin word for border or edge.)

There is controversy over the limbic system because its boundaries have been redefined so many times due to advances in neuroscience. While it is true that limbic structures are closely related to emotion, the brain should be thought of as an integrated whole with emotions flowing throughout it. As we will discuss, mastering our emotions is key to unlocking The Source, especially in the modern world where the powers of emotion and instinct have been sidelined by an overemphasis on logic and analysis. We have become heavily biased in favor of logic when it comes to making decisions and "measuring" success. This is often at the expense of our deepest wants and needs. Just as the physical limbic system has been a subject of controversy, so will be the concept that mastering emotion, and an integrated brain, are far more important to unleashing The Source than logic and the usual symbols of material success. By the end of this book you will know this to be true.

The limbic system processes information coming from our cerebral cortex and sends most of its output of this information to the PFC and hypothalamus, forming a vital hub because it interprets current situations in terms of pattern recognition and "makes sense of them," integrating emotional, logical and intuitive data to help us decide how to respond. This includes everything from the simpler "My baby is crying, maybe she is hungry" to the more complex and emotive "These nagging doubts might herald the end of my relationship." The amygdala are two bundles of cells (one in each hemisphere) which are the seat of our emotional reactions, most particularly the negative emotions of fear and anxiety, and they appear to have a role in attaching behavioral significance to, and creating the response patterns to, any trigger. While longer-term memories are stored further away in the cortex, the neurons in the hippocampus display a high degree of malleability and potential for new growth, which is important in short-term memory and mood.

Deep inside the brain, the hypothalamus receives input from the retina, hormone levels, salt/water balance in the blood as well as body temperature. One of its most important functions is to link the nervous system to the hormonal system via the pituitary gland, intensifying the brain–body connection. The hypothalamus sends information out to the body and has a role in the sleep/wake cycle along with the pineal gland, which releases melatonin as part of our body clock. Interestingly, the famous seventeenth-century philosopher and scientist Descartes believed the pineal gland to be the "principal seat of the soul," and although most of his theories have been disproven, there is still a popular association that the pineal gland corresponds to the Hindu and Taoist concepts of the "third eye" (the mind's eye or inner eye). The third eye signifies the subconscious mind and is said to connect people to their intuition. It can supposedly be made more powerful through yoga, meditation and other spiritual practices such as qigong.

Finally, in the limbic system, the *basal ganglia* are a network of cells in the brainstem that we used to think were primarily to do with voluntary movement—patients with neurodegenerative diseases such as Parkinson's and Huntington's demonstrate changes here—but it turns out they are also key to our levels of motivation and action (both mental and physical). Harnessing the ability of the basal ganglia cells to sustain actions to gain rewards and to liberate us from apathy and inertia—to stay motivated—is another key to unlocking The Source. This is what ensures we keep going to the gym to get in shape, or persevere with studying to put us in a better position on the career ladder.

Brain chemistry

We need a balance of chemicals in the brain to keep our body and brain working in harmony and healthily, and imbalances can have serious effects on our behavior and emotions. In extreme cases dopamine imbalances lead to schizophrenia and serotonin imbalances underlie depression and bipolar disease. Neurotransmitters are the chemicals found on the "root" end of neurons that enable the electrical message to pass to the next target neuron and form pathways in the brain. There are several different types of neurotransmitters in the brain, but the most important ones for The Source are those most closely related to our behavior. Dopamine is the chemical most commonly associated with pleasure and reward (as well as movement). It's part of that craving we get for chocolate or wine and the buzz of falling in love; but unfortunately, it's also the "highs" of drug addiction and consumerism, overeating and shopping beyond our means. Serotonin is known as the "happy hormone," most commonly indicated in the balance of mood and anxiety.

Oxytocin underlies the contractions of childbirth, breastfeeding, cuddling, love, trust and bonding. And we've all heard of endorphins—neurotransmitters that are part of moments of high excitement, an "endorphin rush" as we call it. Perhaps it's winning a race or having great sex, whatever the cause we feel the rush of endorphins as a reaction in moments of exercise, stress, fear or pain as the brain tries to calm us down and limit any perception of pain.

Because the levels of these transmitters and our resultant moods, emotions and drives are absolutely correlated, we can take back the power to moderate our very physiology through how we think and look after our bodies, rather than remaining at the mercy of the levels, supply, quality and flow of these chemicals. The Source will help us to learn how to think abundantly, smile until we are happy, exercise to improve mood, learn to delay gratification, meditate to allay anxiety, and so on.

The key to maximizing The Source is to allow each of the neurotransmitter pathways of the brain to fire up while maintaining a level of balance and counterbalance between them, with constant feedback of information allowing us to adjust the levels of our neurotransmitters and outputs from each pathway. We need to do this to formulate and pursue our goals; make perceptive deductions that encompass *all* of our ways of thinking (particularly emotional *as well as* rational); and to judge risks. Ideally, in this balanced and boosted state we are able to be calm *and* spontaneous, motivated *and* insightful. We have the right ingredients and the best conditions for growth and balance.

The Agile Brain chapters (pages 107–185) and corresponding 4-step program for unlocking the full power of The Source (pages 187–244) see the balance of logic and emotion as a key objective. These two key pathways counterbalance one another and are two sides of the same coin: all of our decisions are biased by emotion, and our emotional impulses in the limbic system are regulated by the PFC's rational thought. We'll explore this in more detail in Part 3: it's the yin and yang in perfect balance and the two poles of brain agility. But when we are stressed our decision-making becomes irrational and erratic, and we lose this sense of balance, either swinging wildly between the two poles of logic and emotion, or staying firmly in one camp, and either refusing to acknowledge our emotions and rationalizing everything or giving in to extremes of emotion and unable to face facts.

Self-care

The demands of modernity conspire to throw our brains into a constant state of overwhelm and stress, so The Source needs our help to maintain its focus and maximize its efficiency. This is where changes to our lifestyle—everything from the foods we eat, number of hours each night we sleep and physical exercise we get—can bring huge incremental gains.

We assume our brain has everything it needs to do its job well and expect it to get on with it. We wouldn't treat our car with such lack of care; we would ensure it was regularly serviced and, if anything seemed out of sorts with it, we would take it to be checked. So why do we tend to assume our brain will maintain its ability to function at its optimum level and be able to prioritize our best interests when we are overtired, eat badly and work in a stressful job without regular breaks? Or when we just "zone out" for large parts of the day?

It feels appropriate as we introduce our amazing brain and its power to direct our future, that we take a quick overview of how and why we need to give it some basic care.

Rest

Sleeping less than the 7–8 hours per night that is optimal for most adults (according to a recent report by America's National Sleep Foundation based on the advice of 18 leading sleep scientists) is not sustainable for 98–99 percent of the human population.[3] Failure to fulfill this quota impacts on a whole host of measures of brain function. Over time, habitual paucity of sleep leads to an elevated risk of everything from Alzheimer's disease to obesity and diabetes. The link between poor sleep and dementia is because the cleansing system of the brain, known as the glymphatic system, takes 7–8 hours to flush toxins out of the brain. These build up over time due to oxidative processes such as stress and alcohol, and potentially lead to the symptoms of dementia. This shows the long-term influence of sleep on The Source, but its immediate impact is also extremely damaging. Lack of sleep has a serious impact on the functioning of The Source, and if you're serious about harnessing your full brainpower, you can't afford to ignore it. A whole night's missed sleep has been proven to impact on IQ.[4]

Sleep deprivation is also linked to increased brain reactivity, which means that responses are more likely to come from the primitive part of the brain, rather than the more logical PFC. A well-rested brain will be able to make better decisions, respond more quickly to stimuli and will have better memory recall than a brain that is low on sleep. When you have had enough sleep, you'll also find it easier to manage your emotions and mood. Sleeping in the side position is the most efficient for allowing the glymphatic system to cleanse the brain, so if I'm having a restless night, I take the opportunity to turn onto my side before I fall back asleep.[5]

There are lots of recordings on YouTube for pre-sleep meditations such as yoga nidra or psychic sleep. A study by the University of Southern California and the University of California found that 58 percent of insomniac participants showed significant improvements in sleep quality with regular meditation. At the end of the study, 91 percent had been able to come off or reduce the dosage of their sleep medication.[6]

Improve the quality of your sleep by:

- Committing to getting a solid 7–9 hours' sleep per night.

- Creating a soothing wind-down pre-sleep routine and avoid your screens for an hour before bed.

- Using a pre-sleep meditation or visualization to help you drop off.

Fuel

Our brain comprises only 2 percent of our body weight but uses up 25–30 percent of what we eat, and it cannot store fuel for later. Research shows that being hungry impacts significantly on decision-making—on big-picture decisions as well as minor ones. Judges, for example, are more likely to grant parole early in the day or just after lunch, when they are more energized and aren't feeling hungry. One study looked at more than 1,000 decisions by Israeli judges and found that prisoners who were one of the first three cases considered after a meal were up to six times more likely to be released as the final three prisoners of the session.[7]

Eating a healthy and balanced diet, rich in protein, with some whole grains (they contain all the essential amino acids that are the building blocks of cells) and "good fats" (such as coconut oil, oily fish and avocado), with vitamin- and mineral-rich vegetables has a huge impact on the brain. On the flipside, too many highly processed foods (such as cakes, cookies and other convenience foods), lots of sugar and too much saturated fat (and trans fat in particular) can be detrimental to your brain, increasing your risk of dementia and a range of mood disorders.

The more we understand about the influence of nutrients on the brain, the more important diet becomes to anyone who wants to take their brain function seriously. Try the following to boost your brain power:

- Eat a teaspoon of coconut oil most days of the week.

- Cut out processed foods and eat more salmon and avocado.

- Cut down on sugary snacks and snack on nuts and seeds instead.

- Up your intake of green leafy vegetables such as spinach and broccoli.

A WORD ON NOOTROPICS

Nootropics are cognitive-enhancing substances that boost brain power—often called "smart drugs." For many years students and businesspeople have used copious amounts of caffeine to work harder for longer and now some take the drugs meant for ADD, dementia or narcolepsy to improve their performance. There is little or no evidence that these drugs enhance cognitive power, but only that they increase wakefulness. As my neuroscience professor friend said: "They're a bit like Viagra . . . they may improve your performance one-off but they won't save your marriage!" They don't "improve" your brain.

Hydrate

The brain is approximately 78 percent water, so it's easy to understand how brain function directly relates to hydration levels. A 1–3 percent decrease in hydration levels can negatively impact our focus, attention and memory. This is why it is essential that every child carries a bottle of water in their school bag, and as adults we should do exactly the same.

We use water to facilitate a number of important bodily functions, such as lubricating joints and carrying nutrients and oxygen to cells. If we don't drink enough water, our body is unable to carry out basic functions and the first areas to be drained of this vital resource are our attention and memory, as our brain does not prioritize them as vital to our survival. But in the modern world, they are. A 2015 study found that dehydration was shown to be akin to driving at the legal alcohol limit in terms of its impact on concentration and reflexes.[8] It found that drivers who just consume a sip of water (25ml) per hour make double the number of mistakes than someone who is properly hydrated. A comparable amount of errors made by an individual was that of someone with a blood alcohol content of 0.08 percent—the current drink-drive limit. A 2013 study by two universities revealed that people who consumed a pint of water before carrying out mental tasks had reaction times that were 14 percent faster than those who did not have a drink.[9] And if you were wondering, we should each be drinking half a liter of water a day for every 30 pounds of our body weight.

If you notice you are thirsty or have dry lips, you are already way more than 3 percent dehydrated. Just like driving a car without topping off the water tank, the brain simply cannot send the chemical and electrical messages it needs to in a dehydrated environment. Ensure that you:

- Observe your thirst levels through an average day. If you find yourself feeling thirsty, you are dehydrated. Aim not to get thirsty by regularly sipping water.

- Invest in a reusable water bottle (BPA-free), and keep it topped off and at hand at all times.

- Swap caffeinated drinks for water or herbal tea, especially if you normally have lots of cups of coffee or tea during the day.

- Eat more hydrating foods such as cucumber and melon.

Oxygenate

Exercise not only energizes our body and brain, causing us to breathe more deeply, which oxygenates cells throughout our body, it has also been found to improve neuroplasticity itself. It counts as one of the factors of what neuroscientists call "environmental enrichment," and research shows that exercise can impact on the survival and integration of cells generated into our neuronal circuitry, by increasing oxygen supply and ensuring there is a "a constant supply of ready-for-action neurons that might either replace old neurons or augment them."[10]

Regular exercise has a host of tangible health benefits for the brain. The combined results of 11 studies shows that regular exercise can reduce the risk of developing dementia by 30 percent.[11] It also makes brains more agile. Those who exercise have better higher brain functions like emotional regulation and flexible thinking, and are better able to quickly switch between tasks.[12]

In a study published in the journal *Neuroscience Letters*, researchers from the University of Texas looked at the impact of high-intensity exercise on a protein called BDNF, short for "brain-derived neurotrophic factor," which means growth of nerve cells.[13] BDNF is involved in brain cell survival and repair, mood regulation and cognitive functions such as learning and memory. Low BDNF levels are associated with a host of mental health disorders, including depression, bipolar disorder and schizophrenia. In the Texas study, all adults who performed a session of high-intensity exercise experienced higher BDNF levels and improvements in cognitive function. But what about how you feel while you're doing it? Believe it or not, when we do exercise that we enjoy we release more BDNF than we do when it feels like a chore. Intention appears to be important in brain activity: wanting to do something, characteristic of an optimistic, abundant attitude, makes it more beneficial.

Walking and other aerobic exercise has been shown to create changes in the hippocampus—the part of our brain that relates to memory, learning and emotional control.[14] The increased plasticity in the hippocampus and possible growth of new cells caused by BDNF—and increase in blood vessels supplying oxygen to that area during aerobic exercise—actually leads to growth in the volume of the hippocampus part of the brain. This also prevents the natural atrophy of brain cells over time, so even a brisk walk is a way of maintaining and future-proofing our brain.

Why not take up table tennis or any sport that involves coordinating multiple factors as well as a social element? This combination of coordination and socializing has been shown to increase brain thickness in the parts of the cortex related to social/emotional welfare.[15] Exercise to build muscle that includes variety and coordination, such as dance, also has brain benefits. And finally, my personal favorite for the mind and body is boxing—it involves cardio, muscle toning and is the best stress reliever I have found in all my own experiments with exercise and mindfulness.

Finally, for many of us living in increasingly polluted cities, air quality is the elephant in the room when it comes to well-being, and a subject that we will all be talking about in the near future. It's one thing we can't control, so it's often easier to ignore its impact. Exercising in polluted areas actually decreases the secretion of BDNF compared to exercising in a clean environment or not at all! When we exercise, we breathe deeply, and choosing to do so at the side of a busy road means we're taking deep lungfuls of highly polluted air filled with toxic microparticles. Monitoring of air quality on a busy central London road concluded that the levels of nitrogen oxide inhaled by pedestrians and motorists were equivalent to smoking four cigarettes per minute—so not promoting the growth or connection of new cells and possibly even inhibiting it.[16]

Here are some tips to bear in mind when you're planning your workout:

- Schedule regular workouts (aim for 30 minutes three times a week minimum) doing something you enjoy. Put them in your diary now so you don't skip them. It could be anything from tennis to dance or swimming.

- Never exercise on busy roads or pavements near traffic if you can avoid it. The air pollution will decrease your levels of BDNF, canceling out the exercise benefits your brain may get.

- Vary the pace of your training, interspersing shorter, fast-paced intervals with longer recovery intervals. This is more beneficial to the brain and BDNF production than endurance exercise at a steady pace.

Clear your environment

Our physical environment plays a major part in maintaining our mood, perspective and stress levels. Take some time to also consider the practical realities of the places where you spend the most time, and assess the impact of this on the functioning of The Source for you. Ask yourself the following questions:

- Is my home calm and happy?

- Is it somewhere I can think clearly?

- Is my workspace somewhere I can be creative and focused?

If the answer to any of these questions is no, you need to think about what practical measures you could take to improve them, such as choosing pleasing textures and aromas, as well as selecting inspiring imagery and objects for your walls and shelves.

A home environment that offers a pleasant sensory experience is one that will help you feel calm and secure; a space to recover from stresses and worries. Although factors such as these are not as significant to our well-being as the amount of sleep we get, they will still impact on our energy, motivation and self-image. Ensuring that you minimize clutter will help you to feel in control. Having said that, everyone has a different tolerance for mess and disorder (my best friend seems perfectly capable of living with a "floordrobe" while my sartorial organization could almost be diagnosed as OCD). Understanding yours and taking measures to ensure you stay within your own healthy range will create a space where your brain isn't assaulted by distracting disorder everywhere you look. The same goes for your office, your desk at work or your desktop on your computer.

Try one or all of the following. I promise the effort will be worth it and you'll feel the benefit instantly:

- Have a major clear-out of clutter at home.

- Try a workspace makeover: filing loose papers and publications away, clearing your computer desktop, choosing some empowering art.

- Delete distracting apps from your phone, and look for ways to clear up your tech habits.

Now that you have identified areas for improvement, turn to the back of your journal and start a to-do list of everything you want to change from now on to ensure you are doing everything you can to support your brain in sleeping and eating well, drinking enough water, exercising regularly and clearing your environment.

All of these lifestyle factors influence our neuroplasticity and brain pathways as our behavior (whether positive or negative) becomes habitual. The more respect we afford to our brain and body by prioritizing sleep, nutritious meals, drinking plenty of water and strengthening our mind and body with exercise and mindfulness, the more positive energy we will have and the easier it is to be in balance within ourselves.

In the following chapter, we'll explore just how flexible and capable of being positively directed The Source is. This is not just about improving the brain physically, but rather, fundamentally changing the way we live our lives.

Chapter 4

Your Malleable Mind: How to Rewire Your Neural Pathways

"The illiterate of the twenty-first century will not be those who can't read and write, but those who cannot learn, unlearn and relearn."

Alvin Toffler

F our years ago, I was told I would soon need reading glasses. I'd noticed I had begun to hold books and my phone further away in order to read the print clearly, and that fiddly little necklace clasps were trickier to fasten. However, my understanding of neuroplasticity—the ability of my brain to adapt and change—meant that when my ophthalmologist said it was "inevitable" my vision would continue to deteriorate further and that it would be "pointless" to resist using reading glasses, I resisted.

I explained to her that I wanted to use my deteriorating vision as a neuroplasticity experiment to find out whether I could slow down or prevent the changes. She was bemused, and said she thought it likely I would start to experience headaches and tired eyes if I didn't use glasses.

My "experiment" was inspired by some reading I had done on the impact of psychological priming on aging. "Psychological priming" is the effect that the mindset of aging has on the physical body—how our thoughts about aging affect our physical abilities. One study explored the impact of the environment lived in on age-related physical and mental decline in older people (the original 1979 study was never published in a peer-reviewed journal, but the findings were outlined in Ellen Langer's book *Counter Clockwise).*[1] In 1979, a group of octogenarians were put into settings that were a mock-up of their lives two decades earlier—with "old-fashioned" furnishings, listening to radio programs from the 1950s and with other visual cues. After only a week living in this "old" life, they experienced improved memory, vision, hearing and even physical strength. Even though these settings were geared against their older, less agile bodies (there were no walking aids allowed if they hadn't been using them 20 years earlier, and reading glasses were taken away), their overall health function improved even in these areas. Having to live their daily lives without the things they had come to rely on over the previous 20 years and inspired by memories of being in their sixties, the brain quickly adapted, giving them a new lease of life. The control group who lived in the same mock settings the following week but only reminisced rather than embodied their younger life also experienced improvements but less so.

There were several crucial differences: the experimental group had to write biographies of themselves in 1959 in the present tense, and all participants sent in photos of their younger selves which were shared with the other participants prior to being on site, and were framed and on view at the venue. The control group only reminisced about the past with the focus on how it wasn't 1959 (despite their surroundings). This latter group wrote their biographies in the past tense and had no photos of themselves from 1959, only the current year. On flexibility and dexterity, the experimental group showed greater improvement, and on intelligence tests the experimental group showed 66 percent improvement over 44 percent in the control group. When strangers were shown before and after photos, they rated the "after" photos of the people in the first group as younger than in the before photos! It was replicated in a BBC program, *The Young Ones*, with aging celebrities, with similarly positive results. Note to self: we do not have to be slaves to our chronological age!

I wondered if I could replicate a similar effect myself by resisting "giving in" to my declining eyesight, and forcing myself to manage reading at a distance that felt slightly uncomfortable rather than moving my phone further away or using glasses. I'm pleased to say that it worked. I haven't experienced headaches and have become totally used to keeping what I have to read at the same distance as before, although this was a conscious effort at first. Not only has my eyesight not deteriorated or even remained the same, it has actually improved a little in the four years since I've been using the technique. I find this hugely empowering and reassuring, and you can do the same.

As my mini-experiment shows, with focused effort and determination, it is possible to avoid or delay some of the supposedly "inevitable" consequences of aging. If I had started wearing glasses instead, the muscles in my eyes would have become accustomed to this and my ocular brain pathways would have quickly adapted to the new adjustment. In short, the brain, being malleable, is capable of reversing a wide range of changes that might otherwise seem inevitable.

Resignation to some of the symptoms of aging, and to any decline in mental or physical function in itself, can become a self-fulfilling prophecy. This is because the brain is resource-sensitive. As we know, it uses 25–30 percent of our energy, so wherever possible its default will be the most efficient (easiest) path, and armed with this knowledge, we can force it not to default. Realistically though, this does not apply to everything because I don't know anyone who has been able to reverse or delay graying hair . . . yet!

The Pathway to a New You

With effort and by keeping our brains in peak physical condition, we can forge fresh ways of thinking, strengthening our higher-level "executive" brain functions (complex decision-making, problem-solving, planning, self-reflection) and learning to master our fright-fight-flight primal brain responses.

People often ask me how long it takes to form a new habit (which is underpinned by a new or altered brain pathway). Of course, it makes a difference how complex the habit is. For instance, it takes a lot longer to improve emotional intelligence than it does to master a new gym routine. But neuroplasticity promises that with dedicated effort, change will come. This principle of neuroplasticity—the power to create new pathways in the subconscious and conscious parts of our brain—underpins all of my work as a coach and it is the key to any deep and lasting shift in our habits and thinking.

It's important not to overcomplicate it. Everyday examples of neuroplasticity are all around us. When a colleague and leadership expert that I teach with at the Massachusetts Institute of Technology (MIT) decided to find out more about the latest neuroscience research that was going on there, she shared the story of meeting one of the neuroscience professors who asked her what she had for lunch the previous Tuesday. As she focused on remembering, then told him the answer, he said, "That's neuroplasticity! You just strengthened the connection for that particular memory simply by recalling it." This may seem like a small thing, but it is a simple example of how we strengthen connections in the brain with every thought or memory.

Try it yourself, right now. Call to mind a day: last Friday, for example, or a memorable day further off: a significant birthday.

Think through it in sequence. What happened? Where were you? Who else was there? How did you feel? Is this a happy or a difficult memory? By recalling it, you have fired up another connection between the neurons in the memory area of the hippocampus deep inside the brain. The more you relive a memory and/or the more intense the emotions associated with that memory, the stronger the connection becomes. This is a result of repetition as well as the intensity of emotion, making it either a fond memory that easily floats to the front of the mind or a dreaded memory that you want to forget but keep reinforcing by mulling over it. Either way, remember the phrase "neurons that fire together, wire together." For better or for worse.

The first step in embarking on the mind-expanding and life-enhancing exercises in *The Source* is to understand that the brain is dynamic, flexible and capable of rebuilding its pathways with dedicated effort. Whenever I hear somebody say, "It's just the way I am" (I hear this a lot when I ask people what's keeping them stuck or limiting their goals), I challenge this belief. It's so important that you fully grasp what neuroplasticity means; in particular, what it will mean for you. It needs to make sense to you personally.

Reclaiming your power

What is the first thing that comes into your mind when you think about what you would like to change about the way your brain works? Imagine what life would be like if you operated from a different paradigm—greater trust, abundance or flexibility. Would you be happier, healthier and have better relationships? Can you see a particular area of your life in which your brain is set into negative habits and pathways? Perhaps you could look back at the statements on pages 10–12 to remind yourself of those that resonated most with you.

If it's helpful, think of your brain as a tangible structure like the hardware of a computer—the keyboard, monitor and drive. Your mind then is the intangible software that you run on this computer. But in this metaphor, you are not a computer that sits on your desk powerless to change. Instead, you are both the coder who upgrades the software to transform the data (your thoughts) and the engineer working behind the scenes to fine-tune the hardware itself (your neurons). You also control the power supply that fuels the computer, with energy determined by the choices you make about what to eat and drink, when and how to exercise and meditate, who to interact with and where and how to live. You are the architect, designer and housekeeper of The Source, with the power to create, maintain and destroy your neural connections. This process is neuroplasticity in action.

Anyone who doubts this power can look to science to provide remarkable examples of neuroplasticity. Neuroplasticity, at its most positive, is the key to self-empowerment. It ensures that with effort, we can overcome deeply entrenched negative behaviors and modes of thinking, including addictive and destructive habits and relationship patterns. I've seen people rehabilitate from the physical effects of strokes and brain tumors, from addictions to drugs and alcohol and eating disorders—and, just as importantly, the more everyday challenges of life such as divorce, heartbreak, bereavement, redundancy, relocation or total career change.

Neuroplasticity also ensures we can achieve forgiveness. Letting go of a past loss or hurt can be the hardest change to make in the brain but often this very pathway is the one that is driving the shame, mistrust and inability to forgive that keeps us stuck. Our brains are constantly evolving, refining and learning in response to everything that we experience—events, emotions and people—and we need to be aware of this and manage what we expose our brains to and how we deal with the impact. We can do this in real time, overwriting past hurts and cleaning up what is present.

The adaptive, regenerative power of the brain is incredible. Whenever we feel trapped by our thoughts or long-established behavioral patterns, it is helpful to remember this. Even some of our most basic "intrinsic" traits can be rewired. Major pathways in the brain can adapt well into adulthood. The "Silver Spring" monkey experiments in the 1950s to 1980s—which became notorious because the monkeys' treatment led to the founding of PETA and high-profile attacks by animal rights activists—demonstrated that monkeys who had their afferent ganglia (the parts of the central nervous system that supply sensation from the arms to the brain) cut, and their dominant arm strapped up, quickly expanded the part of their brain associated with their non-dominant arm as it took over the usual functions of the other arm: feeding and grooming.[2] The results were a landmark in the development of neuroscience, as it was then possible to see that significant "remapping" had occurred inside their brains. This demonstrated that, contrary to what had been previously thought, the adult primate brain *could* change its structure in response to its environment. Being our closest relatives, it was soon shown that this too occurred in the adult human brain.

Edward Taub, the psychologist who led the Silver Spring studies, later went on to use his understanding of neuroplasticity to create a method for rehabilitating stroke victims. "Constraint-induced movement therapy" helped many victims to regain the use of limbs that had been paralyzed for years. This ability of our brains to overcome even seemingly insurmountable challenges (like paralysis) opens up huge possibilities. I use these examples to encourage people: "Look, we really can radically change our brains and therefore ourselves, with effort and persistence."

From the 1990s onwards, neuroplasticity research exploded. In studies that inspired a million "tiger moms," brain scans revealed that playing an instrument leads to great increases in neuroplasticity and new connections all over the brain.[3] The neuronal mass of a number of regions of musicians' brains is far denser than that of non-musicians. Some of this increase appears obviously located in certain areas: brain scans of violinists show that the area of their brain associated with the left (fingering) hand was far denser than the same area in the general population, for example. Other changes show up elsewhere in the brain demonstrating that there are global brain benefits to playing an instrument that are not directly related to the learning itself, like better memory processing and problem-solving skills.

This and the similar global brain function effects of early bilingualism indicates there's a neurological "butterfly effect" at work, with changes in one pathway of the brain triggering changes elsewhere. The positive benefits of neuroplasticity-inducing activities are complex and varied.

Neuroplasticity allows for compensation too, just like in the Silver Spring monkeys and stroke cases mentioned above; studies using neuro-imaging scans of the brains of those who are born deaf show that the areas of the brain usually devoted to hearing are taken over for processing vision.[4] Cases have been recorded of people missing most of one side of the brain, or all of a major region, like the cerebellum. The brain, in all these cases, steps in to compensate in varied and unexpected ways, so that the right side may take over many functions of the left, or the functions of the missing part may be picked up by another region. All of this highlights not only the mystery of the brain and our lack of understanding of it, but its remarkable plasticity and resilience. In this context, the changes most of us are hoping to make are relatively small, which is encouraging!

The Mechanisms of Neuroplasticity

In scientific terms, there are three distinct processes for neuro-plasticity: learning, perfecting and retraining.

Learning

This most obvious form relates to synaptic connection: knitting together stronger links between existing neurons through an increase in the number of synapses (see page 70). This learning is your B+ grade skill—an area you know you have potential in and could be good at if you had enough time to put in the effort required. This could be taking up Spanish which you haven't used much since you were at school, by getting lessons, practicing as much as possible and planning a long vacation to Mexico. In "learning" mode, you would never be as fluent as a native speaker but would be able to hold your own in a conversation and travel well with it.

There appear to be at least two types of modifications that occur in the brain with this kind of neuroplasticity:

1. A change in the internal structure of the neurons, mostly that they develop new synapses at their endings that are able to make more connections with other neurons.

2. An actual increase in the number of connections between neurons so more neurons are connected to each other through these synapses.

Perfecting

Perfecting correlates with a process called myelination—a sort of speeding up of the way the neurons work by starting to coat them in a white, fatty, electrically insulating layer (myelin) that speeds up transmission along them. This maximizes the efficiency of the pathways made up of neurons that are already connected, like insulation that ensures the electricity is maximized and not dissipated.

This tends to occur when you become expert at something, which can be identified if you take easily to something even after years of neglect. This is your A grade skill. You may have a good ear for music and have played the piano and guitar over many years. You then decide to perfect your guitar skills by joining a band and performing regularly. It seems to come to you easily and the more you play, the more your brain adapts.

The best example of "perfecting" is The Knowledge, learned by London taxi drivers, which involves memorizing every single street in the city during training. When undergoing this learning, scientists at University College London have shown that the navigation and memory part of the taxi driver's brain, located in the hippocampus, physically grows in density.[5] However good your sense of direction, this learning requires intense effort. It takes most taxi drivers between one and ten years to crack it but then they are experts. No matter how good our sense of direction, most people will never be as good at this as a London taxi driver.

Retraining

The scientific term for this third process of neuroplasticity is "neurogenesis." It's not as well understood as the other two forms of neuroplasticity, and it occurs far less in the adult brain, relating more to brain change in babies and young children. It involves growing new, mature neurons from embryonic nerve cells which are as yet unformed but have the potential to become neurons and connect up to other existing neurons to form a new pathway not previously in the brain, i.e., developing a new skill you do not have already nor have a natural aptitude for.

This is hard work and time-consuming as it needs to be followed by "learning" and possibly "perfecting." Studies show that in humans, neurogenesis reduces significantly with age. Some studies indicate it barely exists at all in adults.[6] Embryonic neural cells have been seen around the hippocampus where we lay down memories but currently not elsewhere in the adult brain. This makes sense as, in practical terms, trying to train yourself to acquire a new skill that is alien and new is likely to be frustrating and something that only those who have lots of spare time and energy would consider embarking upon. It would be like taking up golf if you've never played before, have poor hand-eye coordination and don't enjoy it. For some people, there would be very little progress before having to give up. For others, you may be able to work towards a mediocre level of competence if you tried really hard, but you would have to ask yourself if your efforts would be better spent elsewhere!

SOPHIE: CUT OFF FROM HER BODY

Some years ago, I was employed to run a resilience program at a law firm where Sophie was a partner. She was an ex-smoker in her fifties, extremely overweight with tired, dull skin, labored movements and low energy. Sophie was on medication for high cholesterol, high blood pressure and diabetes. She explained that her diabetes was poorly controlled and had been getting worse over the last couple of years.

It was clear to me that being a high-achieving workaholic had become part of Sophie's identity and distracted her from the physical issues at hand. As a way of getting her to see this, I asked her to wear a heart rate variability (HRV) monitor for three days and three nights, which monitors sleep, stress levels, physical activity and overall resilience. The HRV monitor works by picking up signals from the nerves around the heart so we are able to see when stress happens and the fright-fight-flight system kicks in. Depending on the heart rate and variability, we can distinguish if the stress is physical or psychological.

The following week I was in shock as the results of the HRV monitor had come up as a complete blank—something I had never seen before. When I shared this information with Sophie, her tone was dismissive: "Oh well, I know I have diabetic neuropathy," she said. I couldn't believe she hadn't mentioned this to me before. Diabetic neuropathy is a form of nerve damage that results from severe, long-term poorly managed diabetes. It means that the ends of the nerves have started to wither away and, affecting the nerves around the heart as was the case with Sophie, is a massive risk factor for cardiovascular disease and ultimately a heart attack. I felt that she needed a reality check. I explained that she had almost every single risk factor for a heart attack—obesity, stress, high cholesterol, high blood pressure, diabetes, a history of smoking—but she felt stuck and had a deeply ingrained denial of her physical state and the consequences of her long-standing poor lifestyle choices.

I made it clear that this was a choice she was making for herself that could have dire consequences for the people that depended on and loved her. I could see that the pathways in her brain that had made her so successful at her job were set on supporting and denying her neglect, and borderline abuse, of her body and health. It didn't seem like she believed she could change, but behind the scenes my warning had hit home as the severity of what I'd said resonated deep in her emotional core. Sophie began to alter the way she thought—the start of retraining as the thought process led to new actions.

The shift in her thinking inspired her to make behavioral changes she had never before felt the motivation to follow through. The next time I saw her she had visibly lost weight and her skin looked less ashen. She said that the day after our conversation she had started walking to work and then taking the stairs instead of the elevator. Over a short period of time she built up to 10,000 steps a day then even started walking to work and back every day (several kilometers each way). She started drinking a green juice every day then totally overhauled her diet.

The learning stage intensified as Sophie's new behaviors built up a momentum of their own once they became habitu-ated and created new pathways in her brain: "I got over the initial pain of breaking myself into new habits," she told me. "I began to enjoy walking and crave healthier foods. I stopped being cavalier about my health and I began to take more pride in my body and my well-being." Sophie cumulated the retraining with learning, and the synaptic connections to support her new, healthier behaviors had been firmly established towards building strong new pathways. The old pathways underlying her negative behavior would have withered in tandem with this new growth, allowing her to overwrite the old with the new.

Neuroplasticity and You

There isn't a universal prescription for implementing structural change in the brain; what works for one person may not be right for you. Dr. Lara Boyd, Director of the Brain Behaviour Laboratory at the University of British Columbia in Canada, has done research that shows how patterns of neuroplasticity vary widely from one person to the next.[7] She describes how the unique neuroplastic characteristics of your brain will be influenced by your genetics. One thing is for sure: as you might expect, creating new neural pathways is hard work. It will feel counterintuitive at first and it's something you need to keep committing and re-committing to. You will slip up, of course, falling back into old ways of thinking and reverting to your habitual pathways. This explains why in the process of learning a new skill—whether it's playing a musical instrument or learning a new language—you may experience the frustrating sense of feeling you've "got it" one week, then find yourself back to square one the next. Brain change happens in phases, and short-term increases in brain chemicals that stimulate connections between neurons aren't the same as long-term structural change that occurs through repeated effort. The repeated effort that then becomes natural behavior is how we develop and sustain habits. These are represented by strong pathways in the brain that are thicker, more connected and possibly better insulated than the next best pathway for that behavior.

Encouraging brain change is taxing physically as well as mentally. The key is to expect this, not to assume it will be easy. When recently I decided to set myself a neuroplasticity challenge to learn a new and entirely different language to any others that I have ever spoken (Danish in my late thirties), I was very aware of the tiredness I felt after about 60 minutes of my 90-minute lessons. I felt tired and then hungry due to the effort required to learn new vocabulary and a new set of linguistic rules that were so different to the rules governing English, Bengali (I grew up bilingual in English and Bengali), French (learned at school from the ages of 9 to 16) or Afrikaans (learned after age 25) that I had studied previously.

It was also fascinating that my experience of learning Danish included the fact that if I were struggling to recall a word, the Afrikaans word would pop into my mind but never the French or Bengali. My friend who is a professor of neuroscience at UCL told me that this was because our childhood and adult languages are stored in different parts of the brain. It was striking, too, that once I had reached a certain tipping point neurologically, I was able to last the duration of my lesson with ease, after two or three months of learning. My brain had done the hard work of "on boarding" all the new rules and processes and I was able to rely on stored knowledge a bit more.

The lesson to take from this is to stick with it when the going gets tough, and to stop wasting time comparing yourself to others or even your own past achievements. Focus only on what you can do now and what you want your future to look like.

Brain scans show that all sorts of activities can induce change in the brain, but three factors in particular have the most impact. Ask yourself how much of each of the following factors you currently have in your life, and how you might be able to introduce more of them:

1. **Novelty**: new experiences such as travel, learning new skills and meeting new people. Novel experiences can even stimulate growth of new neurons. When was the last time you tried something totally new?

2. **Aerobic exercise**: this has been found to increase oxygen-rich blood flow to the brain and allow us to release brain derived neurotrophic factor (BDNF), the endorphin that

allows the growth of new neurons. Do you regularly walk 10,000 steps per day *and* do 150 minutes of aerobic exercise per week?

3. **Emotional stimulation**: the more you experience something and the more intense the emotion associated with it, the more powerful is the effect on the brain. This is why even having shared a traumatic event can be very bonding. We'll delve more into the impact of your emotional responses, both positive and negative, in Chapter 6. In short, though, emotions have a neuro-endocrine effect. For example, sharing laughter with your loved ones has a beneficial effect through the release of the bonding hormone oxytocin which is associated with trust. For similar reasons, break-ups can have extremely negative and long-lasting consequences on your mental health because the high levels of emotions related to shame and sadness correlate to the release of the stress hormone cortisol, which literally locks in connections that loving and trusting someone leads to pain and loss. Can you think of any examples of strong emotions, good or bad, that have locked in strong memories for you?

Neuroplasticity is directed by repetition, for good or ill, so it's worth remembering that negative thinking and addictive behavior can become self-perpetuating, serving to further embed anxiety, depression, obsessive thinking and aggression. Once you fully grasp this fact, you can see why it's so important to harness the power of neuroplasticity to work in your own best interests: to embrace the principle of abundance (see page 25) and the power of metacognition (see page 8). When it comes to the brain it is difficult to unlearn something that has been etched into your brain, and much easier to overwrite unwanted thoughts and behaviors with new desired ones. Of course, these connections ebb and flow in volume and density with use and dis-use. An obvious example is language. If you stop using a language you used to speak, those neurons start to wither away.

What would you like to overwrite in your brain? Which new habits would you like to make, and what new, more helpful pathways could you create in your brain to support your changes? Are there any addictions you need to let go of? Understanding that you can do all of this by using the power of your brain's neuroplasticity is the first step of your journey with The Source.

PART 3

The Agile Brain

Chapter 5

Brain Agility: How to Nimbly Switch Between Different Types of Thinking

"All that we are is the result of what we have thought: it is founded on our thoughts, it is made up of our thoughts."

Gautama Buddha

W e are perfectly capable of accessing more of our brain power more of the time. We don't, because we don't realize how brilliant, flexible and agile our brain can be. Brain agility is the hallmark of optimal functioning in the broadest sense, across all aspects of our lives: work, family, romantic relationships and well-being. An agile brain can:

- Focus intensely and efficiently on one task at a time.

- Think in many different ways about the same situation or problem.

- Switch imperceptibly between these different ways of thinking.

- Fuse ideas from differing cognitive pathways to devise integrated solutions.

- Think in a balanced way, rather than being wedded to one way of thinking (being rigidly logical, for example).

When we're using the full power of The Source, we are unlikely to be drawing on one way of thinking in isolation. Rather, an agile brain is one where each of our neural pathways is adequately developed. Of course, some may be more dominant than others (our strengths or preferences), but brain agility means we are well aware what our strengths and development areas are and that we're able to think in an integrative way, drawing on our whole brain and its resources, playing to its strengths as well as being inclusive about new perspectives.

A Whole-Brain Approach

I use a model of brain agility which describes six ways of thinking that correlate with a simplified version of the neural pathways of the brain:

1. Emotional intelligence: mastering your emotions.

2. Physicality and interoception: knowing yourself inside and out.

3. Gut instinct and intuition: trusting yourself.

4. Motivation: staying resilient to reach your goals.

5. Logic: making good decisions.

6. Creativity: designing your future and ideal life.

Acknowledging the benefits of each of these and then learning how to facilitate their power to work together and in balance— a whole-brain approach—gives us exhilarating control over our brains. This is the antithesis of black-and-white or lack thinking, and is key to developing an abundant and positive mindset.

Just as the different ways in which we think have effects and interactions that skew our perspective, so too do the interactions between the many aspects of our lives. It is not productive to our brain power—The Source—to imagine that if we are having problems with our children at home or have just been through a relationship break-up, this has no effect on our work performance, or indeed that if we lose our job this has no effect on our family relationships or friendships. It is most helpful to understand how this can drain and divert resources and do what we can to buffer the effects.

A useful analogy is that of a gas stove with several burners that we are able to turn up and down in various situations. The maintenance of each flame (brain pathway) as well as maintenance of the gas supply to all of our "burners" is what brain agility is really about. If one of the burners is firing on max for a long time, it will impact on the gas supply available to power our other pathways, and if the full blast continues, this could lead to burnout. It's a helpful way to explain the importance of protecting brain resources, and maintaining balance between the pathways.

So, how does each of these pathways play out day to day? The example below will help to explain what might be going on in each of our neural pathways. We're going to use an example to work out what goes through your mind as you interact with someone. Imagine that you are walking down the street when you see a close friend walking towards you. Read through the examples below, then recall the last time you saw this friend and work through the six pathways making notes for each one:

- Emotions: you might react with a pang of jealousy when you notice her engagement ring flashing in the sun, but mostly you feel a huge amount of fondness, recalling fun memories of times spent together in the past.

- Physicality: you feel a warm feeling in your belly and a skip in your step as you recognize and walk towards her.

- Intuition: when you greet her, you feel deep down the weight of everything she has going on at the moment and you are able to let her know you're there for her.

- Motivation: you work really hard at maintaining your friendship and this has always given you the support you've needed in tough times. You feel motivated to do the same for her.

- Logic: you remember she was applying for a promotion, so you make a mental note to ask her about it and offer your interview tips.

- Creativity: you imagine the future of your friendship. You've visualized her as your bridesmaid when you get married and imagine yourself as godmother to her kids in the future. You understand you are building the foundations for your future bond right now.

This reaction happens in just a matter of seconds and without much conscious thought. In this example, you demonstrate pretty strong intuition and motivation and show emotional self-awareness. There is some work to do on regulating your emotions, but you demonstrate a good amount of physicality and logic, and some creativity. It's normal not to be the same across the board. Most people tend to have two or three pathways that they favor, and two or three that they feel they could draw on under pressure but which they don't consider a strong skill. There is also likely to be one or two pathways that we don't use much, if at all. In the example above, a mixed reaction to a friend may include intense feelings (you feel jealous when you see her ring); this could indicate a difficulty in regulating negative emotion. Similarly, the creative pathway could work harder, helping to turn some of the negative emotional responses into insight and improved self-awareness.

Grouping the neural pathways in this way is a simplification—there is not a singular pathway or discrete group of pathways within the brain governing your decision-making and logical thought, for example. Instead, a dynamic and interconnected labyrinth of pathways ensures your different streams of thought are tangled up together, connecting at various points, like an extremely complex electrical circuit. Within this circuit, some pathways will be far more entrenched than others.

The problem with under-utilized pathways, such as emotional regulation or creative thinking, is that this could indicate that we don't have much capability in that area, that we are ignoring data from that area (filtering information out) because we don't value it, or both. If we ignore data from one pathway—the most common examples I hear are: "I struggle with decisions as I'm not sure what I really want," "I'm not a creative thinker" or "I'm not emotional at work"—then we are behaving as if we don't have that capacity. This impacts on the dominance of our other pathways, which will step in to compensate, distorting the balance of our thinking. To return to the gas stove analogy, if one or two of our burners are blasting at full and the others are left with very little power then The Source isn't working to the best of its overall ability.

Neuroscientist and psychotherapist Daniel Siegel describes this in his book *Mindsight* as the act of "blocking" or "splitting off" parts of your brain.[1] This describes defense mechanisms that we all employ to protect ourselves from difficult or painful feelings. Rather than engage with the messy vulnerability of the primal feelings we all have, we may choose to bury, disconnect or ignore them, or project them elsewhere. Building awareness of this can help reconnect us to our less dominant pathways, and think in a more integrated way.

Much of the focus of my work hinges on efforts to improve people's range of thinking, moving away from a narrow reliance on one or two dominant strengths and towards a more connected, holistic approach. At first, it might feel artificial and laborious to consciously think about a problem or situation from the point of view of each of the six ways of thinking, but using this model eventually trains us to do this seamlessly and simultaneously. Having an agile and integrated brain with free flow of information around it is key to leveraging the full power of The Source.

Below is an exercise designed to examine your brain agility and pathways. Remember that we often have preconceptions about who we are and what we are "good at" and "bad at" due to influences from our childhood. Be careful not to get drawn into this misinformation here and be open and honest with yourself about your brain's pathways.

We will begin by assessing each of our six pathways in turn to get a sense of the agility of our brain, and where our strengths and development areas lie.

Rate your pathways

1. Choose two pages of your journal and draw a circle in the center of the double-page spread and write "The Source" inside it.

2. Draw six arms reaching out from the center, and label each of them with the names of the pathways: emotions; physicality; intuition; motivation; logic; creativity.

3. In order to assess where your strengths and preferences lie, imagine your 100 percent brain resource in the center of the page. Call to mind three examples of recent situations in your personal and work life where you have had to call

on your fullest brain power, like a high-stakes meeting, dealing with a family crisis or making a major life decision.

4. Intuition, creativity and logic are more internal, personal functions. Emotions, physicality and motivation are external as they have an impact on how you relate to others. Bearing this in mind, allocate percentages for how effectively you drew on each of these pathways in the scenarios you have recalled. Are there some pathways that are barely active in your scenarios? If there is a consistent pattern, the lower-scoring pathways are the ones that you may benefit from boosting, as you may be overly relying on the others.

This insight will inform your focus for what to strengthen and practice going forward. They do not have to be equally balanced but you need to feel strong enough in all the pathways as well as knowing what your key strengths are.

It is particularly important to feel strong in terms of managing your emotions, knowing yourself (physicality), trusting your gut (intuition) and staying motivated to create your life as you want it, rather than to be mostly relying on logic to guide you. Perhaps you are more intuitive or creative than you initially believed. Or maybe you have been completely ignoring one of the pathways: blocking its feedback and function. Consider the long-term costs of these patterns and write down your thoughts.

FRED: MULTIPLYING HIS THINKING

Fred worked in a bank and had previously based all his big decisions (even personal ones) on looking at a spreadsheet with the pros and cons listed. He was very logical. He was highly motivated financially, and sometimes let this hunger for success and exceeding financial targets override the less obviously "rewarding" aspects of his job and life. As a result, he made what he knew were some bad decisions, such as ignoring his gut instinct because a deal looked good on paper, or getting caught up in a "herd mentality" rush for investment when an atmosphere of excitement among his peers blinded him to the possibilities of going against the tide and taking a unique stance.

Using my brain agility model with him and working through the pathways, we identified that he needed to work on trusting his gut and thinking more creatively about key decisions. For a while he actually made every single investment decision from the point of view of all six ways of thinking. His intrinsic motivation meant that once he grasped my model, he was determined to make it work for him! It took him about three months until he did not have to do this methodically, and began to trust in his "whole-brain" answer.

Unblock Your Pathways

It's much easier to take on a challenge such as becoming more empathic if it isn't too abstract, and so we need to physically grow a brain pathway by practicing certain behaviors and overwriting the current imbalance with new behaviors. This is far more effective than setting a less tangible objective such as: "Be more understanding of other people's feelings" or "Be more expressive with your own feelings." Often these sorts of demands go hand in hand with explicit or implicit threats such as: "You are going to lose your job unless . . ." or "We are going to break up if you don't . . ." This activates a person's "lack" thinking and this is unlikely to work well as a motivator long-term.

Below are some classic reasons that one of your pathways may be blocked, based on my experiences with patients and clients. See if you recognize any in yourself:

- Emotions: you were brought up in a "big boys don't cry" kind of culture, or your family had very high expressed emotion, shouting and crying a lot, so you find it difficult to control some extremes of emotion in your current life.

- Physicality: you were too small/too tall/weak/overweight/ suffered from acne as you were growing up, or have low self-esteem hence you cower/have poor posture/don't make much eye contact/aren't good at reading body signals and you fear this may have cost you a promotion.

- Intuition: you were harshly blamed or ridiculed for decisions that you made as an adolescent or young adult, so now you feel you can't trust your gut.

- Motivation: you've never had a strong sense of meaning or purpose in life. You've given up looking for a career you enjoy and stick with a stable job that has a good salary.

- Logic: you were told you weren't clever enough to do or be something, so you avoided university and anything that involved exams or needing a good memory.

- Creativity: your teachers or parents said you weren't good at art or music and that all the kids that went into creative industries had always excelled at the arts at school, so you've stuck to safer and more reliable professions and activities.

Raise your awareness of which of these areas may be holding you back. Which of your pathways could be blocked or inaccessible to you because you have clung to an old and unhelpful belief that doesn't serve you? We focus attention on this and offer practical strategies on how to overwrite self-limiting beliefs with new desired behaviors in Part 4, but don't be afraid to start making notes in your journal about what you want from your life and what is holding you back, and begin collecting images that speak to your inner desires that you may want to use later on your action board (see page 203).

OPPOSITES AND THE SOURCE

As I've mentioned, I'm inspired by a range of philosophical and spiritual ideas, and draw on these in my work. In part, the brain agility model harnesses the Chinese idea of balance—yin and yang, light and dark, masculine and feminine, work and life; the notion that polarized forces are required for creation and for life to exist. In true brain agility, The Source is optimized to fire on all cylinders. This leads to well-rounded decisions. From the point of view of neuroscience, this makes absolute sense to me: it is most strongly about balancing logic and emotion.

Opportunity and the Agile Brain

Start to challenge your own ideas of how "whole" and integrated your thinking is. We have all developed preferred ways of thinking and have preconceived ideas about what our weak spots are. Rather than avoid what you consider your weak spots, you could start to think of them as development areas and play around with using different pathways in different scenarios.

You might choose to work through all six ways of thinking, like Fred, or access a modality that you don't usually rely on to trigger new and alternate ways of thinking. The bottom line is that the more of your brain you access, the more power you can unleash from The Source. In the following chapters, we'll work through each of the pathways in turn, looking at the science that underpins them and the practical strategies we can use to maximize their power. As you read through each of the chapters, I'll be encouraging you to think about how freely each pathway is able to flow in your own mind. Is your brain capable of accessing it? Are there any blocks or black spots? What could you do to restore balance and flow?

In terms of making these changes in the brain, it's about developing new habits. It can take anything from 21 to 66 days for a relatively small and obviously tangible change like drinking more water to take hold. But with more complex, less tangible things, such as developing empathy, resilience and confidence, it is better to make a more qualitative assessment than to rely on numbers. Is it making a real difference to your life—improving your relationships or boosting your self-esteem?

Once you fully engage with working through an understanding of each pathway to maximize The Source, you'll have a clear idea of where you need to do most work and what to focus on later in your action board. This may be starting to take shape already as you are collecting images to use on your board. This is the beginning of the road to feeling whole, with the knowledge that you have the tools within yourself to complete the journey that you dream of. Let's get started!

Chapter 6

Emotions: Master Your Feelings

"When dealing with people, let us remember we are not dealing with creatures of logic. We are dealing with creatures of emotion."

Dale Carnegie

We are starting here because emotional mastery is the single most important of all the pathways to work on. This is partly because it is the one people struggle with the most and therefore holds the most potential for change, but also because it is so deep, so fundamental and primal that it holds the highest "X factor" for exponential effects on everything else: our brain–body connection, intuition, motivation and relationships, and our ability to make the best decisions to design our future.

It's also worth remembering that modern life dampens our ability to be fully connected to our emotions in so many ways, not least through social expectations and social media, so to future-proof our careers and lives, honing our emotional intelligence is the most important place to start.

No More Wild Horses

To manifest our full, uniquely human intelligence, we need to learn how to stop acting at the mercy of our feelings, and how to sensitively and accurately read and respond to other people's emotions—at work, in families and in relationships. The balance of logic and emotion is important, as is everything in between, but the traditional or black-and-white idea that "logic is good" and "emotions are bad" is shifting towards the new scientific truth that mastering our emotions holds the key to changing our lives.

Our understanding of the way emotions function has shifted in recent years. Where once emotions were thought of as "wild horses" pulling our minds (the metaphorical cart they are attached to) this way and that, we now understand we have far more control over them than was previously supposed. Brain scans show us what emotional responses look like, how emotions are triggered in the brain and that they can be consciously moderated. Even more positively, modern neuroscience demonstrates that there is plenty we can do to improve our emotional regulation, changing our "internal landscape" for the better, and using the full spectrum of emotions available to enhance our experience of life.

The word emotion comes from the Latin "emotere," which means "energy in motion." Feelings that attach to this energy deliver its nuances and character, and act as a filter through which we experience our lives. Rather than being a passive process, feeling emotions can be reframed as active and generative, though it doesn't always feel this way, particularly when we're in the grip of a powerful base emotion such as anger or excitement. These emotions can seem to "come upon" us involuntarily, but by maximizing The Source, we gain the power to take greater control over our emotions—to act as driver rather than passenger.

In one sense, the idea that emotions "come upon" us contains some truth. Emotions arise in the limbic brain's amygdala, the most primitive part of the brain. Once registered by the amygdala, the brain connects our emotional responses in the current situation to our existing memories (which are stored in the hippocampus). It is then the job of the pre-frontal cortex to decide which of these memories are relevant to recall and what sense to make of our emotions once they have been filtered through the pattern recognition of our past experience. Based on this, our brains use a combination of knowledge (logical thinking) and intuitive, emotional wisdom to interpret and, when required, devise a course of action and behavior in response to what has happened, and the emotions felt.

Sometimes, we can have all our emotions firing at the same time, and achieving any sort of balance in our response may feel like an impossible challenge. When this happens, our brain can be left reeling as intense emotions flood it with a cocktail of contradictory chemicals. In intense jealousy, for example, love, anger and disgust may all be strongly activated at the same time, firing in competition with one another. In such cases, extreme action, such as an aggressive outburst against someone or yourself, may feel like the only way to diffuse these powerful feelings. Everything you learn from The Source and doing the exercises throughout will enable you to react in a way you feel proud of.

Eight Emotion Types

Before we get into how to control our emotions, let's explore the range of emotions available to us, and how to make sense of them. Opposite is a diagram showing the spectrum of eight basic human emotions.

As we learned earlier, all our emotions correlate to levels of certain neurotransmitters. Of the eight primary emotions, the five survival emotions (fear, anger, disgust, shame and sadness)

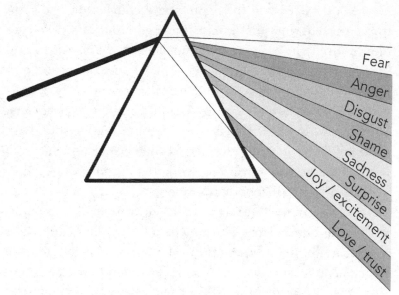

Spectrum of basic human emotions

involve the release of cortisol, the stress hormone. They are likely to be functioning largely below the conscious level, and are all escape/avoidance emotions, generating complex behaviors. It is these feelings that might make you feel like avoiding potentially stressful situations such as public speaking or blind dating. Left unchecked, they can cause you to catastrophize, imagining worst-case scenario outcomes in a way that becomes inhibiting and disempowering. Where they take over the brain's "fight or flight" response, you will feel at their mercy and you may struggle to maintain control of your reactions and keep your cool.

The two attachment emotion spectrums (love/trust and joy/excitement) are mediated by the effects of oxytocin, serotonin and dopamine on the neuron receptors. These activate the reward systems in the brain, and we are primed to want to repeat behaviors that give us these good feelings, like cuddling a loved one or going for a run. This mechanism is helpful in habituating healthy behaviors—you begin to want to go to the gym because you remember it makes you feel great afterwards. However, they can also reinforce negative behaviors. It seems obvious how things like alcohol or relationships with "bad boys" are not good for us and can become an addiction, but eventually even rewarding work and exercise can become addictive, so it's all about moderation.

In between the "survival" and "attachment" spectrums sits surprise. In a category on its own, it is what is known as a "potentiator" emotion that can flip our response state from attachment to survival or vice versa. Noradrenaline intensifies the effects of other neurochemicals and is thought to underlie surprise. This is the feeling we get at the top of a rollercoaster or in a horror movie when we don't know if we will laugh or scream in the next millisecond. Tapping into this pivotal emotion by diverting ourselves away from our usual response can help us deal with a recurring issue in a totally different way; something as simple as taking the advice of a friend who sees problems surprisingly differently to you could flip you out of a survival state and into a mode of abundance and self-awareness. Therapy, coaching or shock tactics like an intervention also aim to achieve this effect by challenging us with a different take on a situation to what we may be telling ourselves.

It is what we expose our brain to that will become ingrained in it, so part of mastering our emotions and staying motivated is about achieving a healthy balance. We need the full gamut of emotions in our life but not too much or too little of any of them.

You and Your Emotions

We all grow up with models for how to relate to other people, express ourselves, give and receive love and handle disagreements. As we move through life, these "imprints" we have tend to get projected onto other situations and relationships in our lives. This powerful unconscious process is one that's worth exploring as it can have a profound influence on the relationships we choose, the way we view ourselves and the way we think and behave.

The emotional profile of your family will have a big impact on how you manage and express your emotions. Those from high expressed emotion families, where there is a lot of charged vocalizing—noisy arguments, shouting and crying—may find it difficult in a relationship with somebody who internalizes more and is very controlled in the way they express themselves. They may also find it difficult to rein in this learned behavior during heated work disagreements.

A short quiz

Ask yourself the following five questions and take a moment or two to ponder the answers, and make notes in your journal if you wish:

1. What was the emotional style in your family growing up? How were disagreements and difficult conversations handled?

2. How in touch are you with your emotional responses?

3. How easy do you find it to regulate your emotions, disengaging from overwhelming feelings like rage and fear, and pulling yourself back to the moment?

4. How easy do you find it to build rapport with somebody new?

5. How often do you experience a feeling of emotional resonance when you are talking to somebody—a mutual feeling of understanding and connection?

Start by building a picture of where your emotional strengths lie and your areas for development. Does your ability to control your emotions change with different people? How do your emotions shift under stress? Are you able to hold it together at work but feel short-tempered when you get home? Are you different on or after a relaxing holiday?

EMOTIONALLY INTELLIGENT? OR JUST EMOTIONAL?

There's a big difference between being emotionally intelligent and being sensitive. Often people who describe themselves as "sensitive" may be so to their own emotions but insensitive (to the point of clueless) when it comes to other people's feelings. It's worth noting that this isn't always a constant. There are some instances where otherwise emotionally intelligent people can slip into egotistical emotional preoccupation that blinds them to the emotional impact they have on the people around them (this is most likely to be acute at times of turmoil such as a divorce or midlife crisis). In my experience both men and women are equally likely to fall into this camp, although it often takes some deep thought and humility to recognize if this applies to you.

NICOLA: "NOT A FEELINGS PERSON"

A woman I worked with—Nicola—described herself as "not a feelings person" and was convinced there was nothing I could do to help her change this. She was a manager at a restaurant, in her mid-thirties, and she had a reputation for giving brutal feedback without realizing the impact it had. She never joined in with the staff birthday tea and cake tradition, preferring to stay at her desk working rather than venture out to chat to colleagues and find out more about their personal lives. Thus, she didn't have the best working relationships she could. I explained: "You can't think yourself into feeling any more than you can feel yourself into thinking." Nicola looked nonplussed and a bit cross at this, so we started with me helping her with some of the emails she had to write to restaurant staff she had difficult relationships with. I advised her to replace half the "I think" sentences with "I feel." She found this easier to do in writing than she did when speaking and it built up her comfort with the vocabulary.

Nicola had young children and, despite a demanding job, devoted herself to having great relationships with them and their au pair. I asked her if she could apply some of her parenting values to nurturing the waiting staff—not necessarily the same behaviors but the values behind them. This struck a chord, and she was increasingly able to offer more of a nurturing approach to her team over time.

After only a few weeks of practicing emotional attunement—things like listening without interrupting, giving good eye contact, paying attention to people rather than multitasking, and using words like "I feel," "I trust" or "I would love" rather than "I think," "I've decided," "I want"—Nicola reported radically different relationships with her staff.

Consider what resonated with you here and whether a similar approach could help you in one area of your life. Could you step back and do things a bit differently at home or at work? Are there some words that aren't representing the best you or new words that you could incorporate into your vocabulary?

A toolkit at our disposal

The power to control our emotional landscape really is in our hands. Our emotions themselves are tools we can use to formulate our responses. I heard Harvard psychology professor Ellen Langer speak at a conference years ago describing emotions as being more like the ingredients available in a chef's pantry than untamed forces of nature to which we are at the mercy. We can choose to act on our emotions as we would choose ingredients in a pantry. Our brain is engaged in this process the whole time, selecting the "ingredients" to combine to form a response to the situation we find ourselves in: a pinch of surprise, a splash of excitement, a shake of fear.

The way I put this is that if somebody upsets us, we could either make scrambled eggs or bake a cake. Our response is very much within our control. Of course, some people will be more adept at recognizing this than others. People who feel at the mercy of their emotions are like zombie chefs, running on an autopilot that drives behavior without them even being aware of it.

A strong ability to regulate emotions is something we need to work at. Practices such as mindfulness help, building and expanding the pause between our thought and our response to it.

RETHINKING "AMYGDALA HIJACK"

There is a concept known as "amygdala hijack" which was popularized by Daniel Goleman in his 1996 book *Emotional Intelligence*.[1] He described a state where you are so overcome by strong emotions—usually fear or anger—that you are "hijacked" by them, powerless to control your resultant thoughts and actions. The science has moved on in the last few decades and we now know that although it may be harder to regulate strong emotions in this state, it is still within our power to recognize, manage and improve our behavior. When I've challenged this belief in the workplace, there is not one person who has failed to go on to better control their temper. It's interesting how what we believe affects how we behave—even unconsciously. Throughout this book, you will be given the opportunity to disrupt yourself by challenging what you have previously held as true—that is not serving you—to create a different future.

Building emotional literacy

Learning about emotions and their impact on us can help us become more emotionally literate, noting and labeling emotions in our mind as they arise. Research on those who regularly meditate shows they have greater emotional literacy and control, and more stable moods than those who don't.[2] A study at Yale found that regular meditation reduced rumination: a type of thinking that correlates with reduced happiness.[3]

Begin by looking at the spectrum (page 125), identifying those emotions you access regularly and those you don't feel have such a big part in your life. It is likely that there will be some emotions you have blocked off or don't have words to articulate. Each of these emotions will affect us differently, but noting your emotions as they occur will help you detach from them and feel more in control. Strong feelings are less likely to engulf you when you maintain a measure of objectivity. Try saying to yourself "sadness" or "anger" as the feelings arise; it may sound like a small thing, but it's surprising how dramatic the impact can be.

I've worked with several clients who struggled to contain their temper. These men and women would go red in the face and scream and shout until their staff was reduced to tears. One man even dissolved into tears himself in front of his junior work colleagues. The incredible thing was that my clients reported to me that some of the time they did not know they were losing it—only after the event when their family and workmates complained did they began to wonder what had happened. This was my cue to dismiss the "amygdala hijack" myth and ask them to agree that their behavior was totally unacceptable and must not happen again. It felt like I was back in the child psychiatry clinic at times!

All of them could see that it had to stop but maintained they would struggle to stop a behavior that they were at best only partially aware of. I asked them to note the next time it happened and afterwards reflect on what they could have done differently. Then on the following occasion they were to make efforts to recognize the emotional overreaction in the situation even if they could not then stop it in the moment. The very next time, I asked them to stop the behavior as soon as they noticed it was starting. They then had to learn to recognize the early warning signs of a "rage" and walk away or practice the STOP method.

Learn how to STOP

I used this exercise when I was working as a child psychiatrist. It's a technique that is often used by family therapists with children who get into uncontrollable rages. I used it again more recently with executive coaching clients.

Close your eyes and allow yourself to feel what it's like when you're overwhelmed with anger. Remember something that makes you angry and allow it to fill your whole body. Feel the anger on your skin, in your chest, your mouth, your muscles and your mind. Once you feel full of it, imagine holding up a big, red STOP sign in your mind and allowing the feeling to dissipate completely, relax your muscles and let the angry feeling leave you. Practice this until you feel you can use it in real-life scenarios to stay calm.

When you start to use physical exercise, a mindfulness practice such as yoga or meditation or the STOP exercise to deal with explosive emotions, you'll find you naturally begin to phase them out as an option over time. However, sometimes the only thing to do when you feel emotionally overwhelmed is to go to bed and start afresh once your emotions have subsided!

A new paradigm

We are all creatures of emotion, whether we acknowledge this or not. Every single decision that we make is biased by emotion. By developing our emotional intelligence, we can maintain balance as our default setting more of the time. Inevitably there will be times of stress that push our minds into lack thinking and survival mode, but the quicker we are at recognizing this and acting to restore the balance, the better able we are to avoid potentially damaging pitfalls.

Intuitively, we will be able to see the way that these pitfalls tend to manifest for us. We may find ourselves worn out by years of feeling overwhelmed with emotional highs and lows we could have avoided in hindsight. Or perhaps we tend to do the opposite, to suppress emotions and overthink everything, then feel in retrospect that we've lost out as a result of our refusal to trust our gut or heart. Any sense of having failed to make good decisions we "should" have is an indication that it's time to embrace this truth: mastering our emotions is our only hope of fully unleashing The Source. It's no good trying to think your way out of your emotions intellectually, but equally the opposite view (regarding yourself as the victim of emotional "weather") is equally unhelpful. Emotions make us who we are and govern our entire experience of the world and life itself.

Can you think of an example of where you have buried certain emotions? And an example of where you dealt with a tricky personal situation really compassionately? Improving our emotional "weather," through better self-care, mindfulness and the practical exercises in Part 4, will help us to live from a more abundant perspective, where we take back power over our emotions rather than letting them rule us, or fearing and burying them.

Next, we'll explore the relationship between our mind and body, and learn how to strengthen it in order to better synthesize our physical, as well as emotional, state.

Chapter 7

Physicality: Know Yourself

"Your body hears everything your mind says."

Naomi Judd

I n a generic sense, building a connection between our mind and body supports all aspects of our self-care: a body that feels good tells us we are looking after it well; a body that doesn't clearly needs some extra attention and support. Body image and "feeling comfortable in our own skin" are also linked with this. Our bodies carry the story of our health behaviors: the triceps that pull us through the water in the pool, the tilt of our chin when we are feeling confident and at ease, the shoulders that hunch at the end of a week at our desk then drop and relax after our yoga class. All have their story to tell. Work through the body scan below and ask yourself what story your body tells.

Body scan

Do this once a day for a week. Note down how you feel afterwards each time. What did you notice? How did you feel?

- Sit comfortably upright in a chair, with your hands resting on your lap and your feet on the floor. Ideally, remove your shoes to feel the floor beneath the soles of your feet. Make sure you are not crossing your legs or arms. Close your eyes and relax.

- **Bring your attention into your body.** Feel whether it's relaxed or tense, surrounded by space or confined by surrounding walls, people or clothing. As you sit, notice the weight of your body, the feeling of yourself resting in your chair, or the floor. Note the feeling of weight where your legs and bottom are in contact with the surface you rest on.

- **Take a few deep breaths,** inhaling slowly through your nose for a slow count of four, and exhaling more slowly still. As you inhale, focus on the oxygen enlivening your body and, each time you exhale, feel yourself relaxing more fully.

- **Turn your attention to your feet on the floor** and focus on the sensations of your feet touching the floor: the weight and pressure, any vibration, the temperature. Lengthen your toes. Move up your body, bringing your attention again to your legs resting against the chair. Note any pressure, pulsing, heaviness or lightness.

- **Notice your back against the chair,** moving up from the base of your spine to its center.

- **Now bring your attention into your stomach area.** If your stomach is tense or tight, let it soften. Take a breath.

- **Notice your hands.** Are your hands tense or clenched? Let them be soft and see if you can isolate your focus to feel each of your fingers in turn.

- **Notice your arms.** Feel any sensation in your arms. Let your shoulders relax without forcing them downwards.

- **Move your attention upwards towards your neck and throat.** Tilt your chin down a little to lengthen your spine.

137

Relax. Soften your jaw. If your tongue is stuck to the roof of your mouth, let it go. Let your face, facial muscles and even your eyeballs be soft and at ease.

- **Send your attention to your crown.** Allow it to rest there, feeling a long line from the crown of your head to the bottom of your spine, then expand your focus to take in your whole body. Feel it as a unified, connected, breathing entity.

- **Breathe deeply** for a count of three, then open your eyes.

Note down how you felt. Were there areas of tension? Did one side of your body feel more relaxed than the other? Were you able to relax when you consciously tried to do so?

Interoception

We are all familiar with the primary senses through which our experience of the world is mediated: sight, sound, taste, smell and touch. "Interoception" is different: it is the lesser-known sense that helps us feel and understand what's going on inside our body. It helps us "read" our own body and process its signals, from hunger, thirst and body temperature to our heart rate or the state of our digestion.

Children who struggle with the interoceptive sense may have trouble knowing when they feel hungry, hot, cold or thirsty. Take a moment now to consider how you know you are full at the end of a meal, or when you need to go to the bathroom. Most people would be able to "read" these signals accurately, but there are more subtle examples of these feelings at work all the time that are more likely to go unnoticed. Knowing when to take a break at work as you feel your energy starting to flag or when to leave the room during a disagreement to avoid shouting at your partner and saying something you're likely to regret are two everyday examples.

Life experience impinges on interoceptive ability. Your capacity to correctly decode physical clues, like loss of libido or appetite, is influenced by the way you have learned either to take note of or ignore different types of feeling.[1] Your family attitude to health, emotions and well-being will have an influence on this. I have noticed with people whose parents had a "stiff upper lip" approach and told them to "get on with it," dismissing concerns with "You're fine" or saying "Don't be so silly" when the child had an emotional reaction, may find it hard to read their body's signs and anything that signals their vulnerability—physical or mental. Unlocking the connection between our body and mind may be challenging if you have taught yourself to override your body's signals to prioritize other things.

Failing to spot and respond to the internal clues that our bodies are communicating to us is an example of us zoning out from the internal state of our body. Our brain is involved in a constant process of integrating signals relayed from our body into specific neural sub-regions—such as the brainstem, thalamus, insula, somatosensory and anterior cingulate cortex—allowing for detailed feedback on the physiological state of the body, from the simple "I'm too hot" to the general and complex "I feel well" or "I feel tense." This process is important for maintaining conditions in the body such as temperature and blood pressure and, potentially, aiding in self-awareness, as the better we get at tuning into these signals, the stronger our brain–body connection gets.

Interoceptive signals are transmitted to the brain via multiple neural pathways related to our heart, blood, lungs and

skin (plus other physiological systems such as the gastrointes-
tinal, genitourinary, endocrine and immune systems). If you
are highly attuned already, then your immune interoception
will be able to tell you when you are coming down with a
cold, sometimes days before you have obvious symptoms. You
know that barely perceptible feeling you get a few days before
you get an illness, when you are pretty sure what's coming but
don't have any obvious symptoms? Where once we thought
this came down to pure intuition, we now understand that it's
interoception at work. Our bodies send out a range of micro-
signals that tell us our immune system is struggling, whether
that's a slightly elevated pulse rate, a foggy headed feeling or
a tingle at the back of our throat. Our brain registers all of
this and warns us something is up if we are able to take note
of it. It is then up to us to take some extra supplements, drink
plenty of fluids and have an early night, or ignore it and push
on through.

Many of us will have suffered from "vacation sickness" and
found ourselves laid up with a bad cold or tummy bug on the
first day of our planned leave. This is a big warning sign that
we are ignoring, or are unable to read, messages from our body
to our brain, and our suppressed immune system is getting its
own back when our defenses are lowered. It can work the other
way too, with stress and anxiety causing psychosomatic pain
that does not have a physical pathology behind it. To recognize
this, I sometimes find it helpful to almost have a conversation
with myself where I check "Am I unwell or in pain for a reason,
or could this be stress?"

Interoception uses physical pathways in the brain and body
in much the same way as our five senses have nerves that
connect up the eyes, ears, tongue, and so on, to the parts of
the brain that process vision, hearing and taste. Although it

is a more complex and mysterious sense that will be further revealed by more research in the lab, it is a sense that we can start exploring to improve our own brain–body connection, our well-being and the potential of The Source.[2] This is why learning to do a body scan exercise is crucial to our personal development.

ANDY'S TRANSFORMATION

I was working with Andy, a 36-year-old man who worked in the media and had developed adult asthma after a bout of pneumonia. He was pale and ill-looking, and he told me that at one point he could not walk from one side of the building to the other without becoming out of breath. He worked very long hours and also spent three hours a day commuting as his family lived out in the country in the dream home he had been able to purchase with the proceeds from his shares in the company. He was now locked in to working in that environment for three years in what he called "golden handcuffs." When I asked him what would make him stop and reassess his situation he said "another hospital admission," and burst into tears.

Every time I talked to Andy, his face would light up when he talked about his past hobby: competitive distance cycling. Now, his packed work schedule and lengthy train commute meant there was no way he could fit it in, and weekends were sacred family time, so he didn't want to spend hours away from his children out training. I suggested that Andy could at least make his money work for him by renting a small flat near his office where he could live during the week, but he immediately said that he couldn't possibly be away from his wife and small children all week.

About six weeks later I saw him again and he was transformed! His skin had lost its pallor, he had a spring in his step and an air of serenity about him. When I asked what had changed he said he had gone home that night and mentioned my suggestion to his wife, and she had immediately insisted that he rent a flat in London and that she would make it work at home. He then told me that he used the extra time in London to stay on top of his work so that he could leave early every Friday, and that every morning he had started going out for a long bike ride before getting to the office. He had also returned to his meditation practice. He had lived in India for a year at the age of 18 and knew more about various forms of meditation than I did! Andy had realigned his body and mind and found a way to live according to what was really important to him, rather than old choices about what he felt his life should look like. By finally listening to his body, he reclaimed his passions as well as his health.

Learning to "read" your body

So how can we improve our interoception? Is it something we can work on, or is it something we have innately? Interoception is often a hard concept to grasp, unless you do yoga or participate in a sport that requires bio-feedback from the body. The more grounded and present you are in your body, the stronger your interoceptive sense is likely to be, but it is absolutely something you can build on.

It's easy to walk down the street and not notice many of the details. Even an area that we have walked around in for many years may feel difficult to reproduce in our imagination if we haven't been paying special attention. It's exactly the same with interoception. Instead of disregarding that twitchy eyelid, those restless legs or the butterflies in your stomach, start to be more curious about what they might mean. Some will have a scientific explanation, for example muscle spasms or migraines can be related to low magnesium levels, but some will have a unique meaning to us and it is down to us to decode them. I have a friend that has always said "By the time I get a mouth ulcer, I know that I have pushed myself too hard and my body is not getting enough nutrients to keep going." I have taken note of that for myself. A cousin said that his stress always shows up as a feeling of toxins built up in knots in his shoulders. I recognize that one too!

Learning to read my body is something I have consciously worked on and improved myself. I'm now adept at decoding my body's hidden signals, and often know days before I get ill—I become acutely aware that something is out of kilter in my body. If it is a slightly scratchy throat or distant earache, I will have lots of warm drinks with manuka honey, lemon and ginger, go to bed early and do some restorative yoga that balances out the neuroendocrine system and helps to restore the adrenal glands which drain our immunity when we're burned out. As a result, I'm often able to stave off what might otherwise have been inevitable.

I believe that each of us has the innate capability to know our bodies and communicate this with our brain to make better choices. For example, if your lips tingle when you eat shrimp or you get bloated after a pizza, you probably have a sensitivity to crustaceans or an intolerance to wheat or gluten. What signs can you think of that indicate stress, dis-ease or being out of sorts for you? Armed with this knowledge, you will be better able to implement self-care strategies quickly, and less likely to push through when your body is telling you to stop.

We can all be overwhelmed by psychological issues and it is important to understand what they are and how they can be managed. But sometimes, if you search inside yourself, you may sense a physical root cause or hormonal imbalance that can be corrected. This could be anything from having a snack when your blood sugar falls low and clouds your judgment, to a vitamin deficiency, all the way up to pre-empting a potentially more serious health scare.

YASMIN: OUT OF TUNE

A shop assistant, Yasmin, discovered a cancerous mole that she would not even have considered looking out for until she watched a morning TV program about self-examination and recalled me suggesting she pay more attention to what her body is telling her.

Though she was one of the lucky ones, this health scare catapulted her need to understand her brain–body connection. Yasmin has subsequently experienced many benefits to her health including weight loss, increased calmness and a desire to try new things following a change in diet, an increase in exercise and a more mindful lifestyle. The whole experience has revolutionized her attitude to self-care and noticing her body.

But you should not have to wait for something life-threatening to happen. Commit to tuning into your body now rather than later. The body scan exercise on page 136 is a good place to start. A food, mood and bowels diary is another way. For a week, note down each day what you eat, what sort of mood you are in (on a scale from one to five), and how many times you go to the bathroom to open your bowels. This simple commitment can lead to some unexpected insights about how your body works, what boosts it and what slows it down.

A window to your emotional state

While interoception primarily relates to physical awareness—talking about it reminds me of past patients willfully ignoring their persistent chest pain and plowing on with their high-stress, unhealthy lifestyle regardless—it also links to more general emotional self-awareness. After all, physical symptoms are often the best (and earliest) indicators of our psychological state, usually before intellectual thought or logic has had a chance to intervene and interpret.

People who find it difficult to "feel their feelings" are likely to struggle in all sorts of ways with relationships, both personally and professionally. This could show up as anything from ignoring a headache to not realizing when they have alienated a colleague or upset a partner. They may have powerful defenses against their own aches and pains or feelings of sadness, as well as inappropriate reactions to the emotions of others, even those closest to them: denial, passive aggression, black humor or escapist addictions. Studies have shown that our tolerance for physical pain directly correlates with our ability to withstand stressful emotions—our pain thresholds are interlinked—and this can pan out in many arenas of life.[3] Have you ever thought about how you respond to a bump or a graze, or how you would feel about acupuncture, for example? And is there any correlation between this and how stoical or reactive you are to stress or disappointment?

You may have experienced finding it challenging to assert yourself, to know what you feel, think or want, and may find it almost impossible to verbalize this. People often say, "I'm not heard" or they studiously avoid confrontation or difficult conversations, not realizing that "healthy conflict" can actually solve a lot of issues that would otherwise fester and become bigger than they need to. Is there an area of your life where you need to speak up or sort out a misunderstanding? Ask yourself what the possible outcome will be if you do this, and what it will be if you don't.

Moving Forward

There is a growing scientific consensus that an effective brain–body connection is also integral to executive function—higher level thinking in the brain. This makes absolute sense, as interoception draws on a combined awareness of your other senses to build a holistic picture of how you feel at any given moment. High interoceptive awareness has been found to correlate with mental and physical well-being. The authors of a 2017 study concluded there were many similarities between the well-being of those who were naturally disposed towards a mindful perspective, and those who had high interoception.[4] This is fast emerging as the factor that underpins self-awareness and physical resilience and energy.

In my brain agility model, all of the pathways rely on a good supply of energy. Energy is essential to maintain our ability to stay self-aware and manage our emotions, keep motivated when we are tired and distracted, rely on our intuition when we don't know who to trust, bring out-of-the-box thinking into our life when we are in a rut, and decide on and stick to our life goals when the going gets tough. All of this begins with the body. On a good day, with the energy of our mind and body aligned, we can achieve what seems impossible on a day we haven't slept well or had time to eat properly.

Can you think of an example where you've ignored your physical state to your detriment? When was the last time you really took good care of yourself, inside and out? How regularly do you do this?

Next, we'll tie together emotions and physicality to access our intuition: the wisdom that resides in our primal brain and gut neurons.

Chapter 8

Intuition: Trust Your Gut

"There is a voice that doesn't use words. Listen."

Rumi

How good are you at acknowledging your intuition? Do you pay attention to it, or carry on regardless, ignoring your nagging suspicions? Do you believe in the value of "gut feeling"? People are often skeptical when I first introduce the importance of intuition in a business context because it seems "illogical," but it is key to everything from good decision-making to improved self-awareness. Here we will be discussing both the importance of honing your intuition to improve your life, as well as how healthy gut flora contributes to positive thinking.

The Gut–Brain Connection

Earlier this year, I had lunch with a former client. I hadn't seen him for years, and in the interim I had met my new husband and got married. "What happened to you?" he asked. "You've got your sparkle back!" When I asked what he meant he said that he had sensed I wasn't in a great place when we had been working together, about six years earlier. "It was like you were at 60 percent, but now you're at full power."

At the time, we had never discussed my personal life, so this feeling was purely a hunch. I told him his intuition was right—things had been tough for me in my personal life when we had been working together. Although I didn't think anything in my outward appearance signaled this, my energy clearly did.

We have these "gut feelings" about one another all the time. Our intuition helps us "sense" the truth of things, tuning into energy that our conscious thoughts may not be aware of. Recently, it has become trendy to talk about the gut as a "second brain," but I think this is misleading and unhelpful. The gut isn't a second brain; it houses the enteric nervous system which is one of the main divisions of the body's autonomic nervous system, and it works unconsciously (in much the same way you breathe and your heart beats without any intervention from your conscious brain).

The gut is a discrete, but connected, system that links back to the brain in a number of ways, including our inner voice about our closest relationships.

The new science of trusting your gut

The connection between the gut and the brain—the subject of scientific study from the mid-nineteenth century—has caused considerable debate. Neurobiological research has revealed a complex communication system between the sheaths of the millions of neurons embedded in the gut walls and the limbic brain that is integral to our decision-making.[1] Scans show these pathways clearly. As we know, the limbic brain is responsible for the experience and expression of emotion. It is where our habits and behavioral patterns are stored; so, the gut–brain system helps govern not only the healthy functioning of our digestive system but also complex brain functions including motivation and access to our deeper wisdom.

Alongside this emotional link is the physical health of the gut. Self-care factors—from diet and supplementation to stress management—that all impact on the well-being of our digestive system, have an impact on our intuition. For instance, there is evidence that something as simple as taking a good-quality probiotic for one month to rebalance gut flora reduces negative thinking. In a study in the Netherlands, probiotic supplementation was found to reduce "cognitive reactivity" to low mood.[2] I take probiotics whenever I travel as the gut microbiome is affected by jet lag too. It is an easy thing to incorporate, and knowing that it also has this general benefit on abundant thinking is an added bonus. Just as our thinking is slower and harder when we have a cold or are in pain, so a depleted, inflamed or struggling gut will cloud intuition. Is this something you have thought about and made a connection with for yourself? And do you care for your gut health in a proactive way? If not, there are potential quick wins for you here. Cutting down on red meat, processed and high-sugar foods will help to reduce inflammation in the gut. Finding out if you are intolerant to gluten or lactose, eating probiotic foods, like kefir, sauerkraut or kimchi, and taking probiotic supplements will also help.

Take a high-quality probiotic every day for the month to come. Quality brands should contain over 50 million strains of good bacteria. (The water-based ones or capsules are far better than the yogurt drinks as they get to the small intestine and are not broken down by stomach acid.) Take note of any effects taking probiotics has on the quality of your thinking.

One thing is now a scientific certainty: the gut–brain link is far from a mystical "sixth sense" and I share the research just mentioned with my skeptical clients when they are tempted to dismiss intuition. Once they hear about the science, they understand that it's certainly worth doing what they can to improve their gut–brain connection and learn how to better listen to their inner voice. Often these people have digestive problems as a result of excessive travel, poor diet, lack of exercise and dehydration. They have poor self-care and still expect themselves to function, both mentally and physically, like a well-oiled machine. One of my first challenges when I start working with them is to persuade them they are short-changing themselves, and that this physical neglect has profound neurobiological consequences.

There is also a growing body of research that links the gut microbiome to our immune system, as the quality of immune cells that are produced in the bone marrow are related to the quality and variety of bacteria in the gut.[3] We don't know enough about this yet but it is an exciting area of research that points to being pivotal in terms of understanding the interconnectedness of our immunity, resilience and optimal brain performance.

Gut and mood

There are other important gut–brain facts to take note of: your gut produces neurotransmitters—a staggering 90 percent of the serotonin that works primarily in your brain is produced in the gut. Serotonin acts in a variety of ways: in the brain as a "happy" hormone that helps regulate mood, and in the gut as a paracrine-signaling molecule, which means it induces change in close-by cells in the body. This action has been shown to impact on weight gain by helping to regulate insulin production.

Self-care factors, from exercising enough to eating a balanced diet, are associated with the healthy production of serotonin in the gut (and the added benefits of this on mood).

Self-care is also key to keeping our stress symptoms under control, and research shows that stress levels are constantly being communicated by the brain to the gut.[4] The brain lets the gut know via sympathetic nerve fibers when it is stressed. The gut responds to this by using less energy for digestion, decreasing the blood supply. When stress is constant, it puts a negative drain on the gut's resources and results in a range of symptoms, from change in appetite to bloating and diarrhea or constipation, or worse. Over time, the gut walls can weaken and the immune cells secrete large amounts of signal substances that lower the stress threshold in our body and brain. Chronic negative emotions are perpetuated in this way, and an unwell gut may mean we feel the effects of stress long after the cause of that stress has disappeared.

Being in tune with this and recognizing the symptoms can help us to monitor our stress levels, keep our immunity optimal and hone our intuition. This could mean anything from knowing that you need to rest and recuperate after a busy project at work or a lot of socializing, all the way up to something as serious as a brain injury that could all too easily be explained away as a psychological disorder if you didn't know better.

JACQUELINE: A LESSON I'LL NEVER FORGET

The danger of interpreting symptoms that appear to be psychological without taking the physical body into account is something I've had to be mindful of as a doctor. When I was a psychiatrist in Bermuda, I had a new patient appointment with a young lady named Jacqueline, who came along with her friend who was a trainee psychologist. She had experienced a severe personality change over the previous few weeks with extremes of emotion and uncharacteristic behavior, being tearful and demanding when she had always been a timid introvert. What was striking was how adamant her friend was that she must get admitted to the psychiatric hospital that day. They had tried before and been sent away and enough was enough—her friends and family could cope with her no more. But something did not fit and I explained that I was unable to admit her to a psychiatric ward and they should return to the emergency department. They left very disgruntled.

Unusually I got a phone call a couple of hours later from the emergency department—these are busy people who do not normally have time for routine follow-up chats. I assumed they must be trying to persuade me to admit Jacqueline back to my hospital. But that wasn't the case. It turned out that she had suffered a small head injury three weeks prior and a pool of blood had accumulated inside her skull and started to press on her brain (a subdural hematoma) causing the personality change. The emergency doctor told me that if I had admitted her to a psychiatric ward and the hospital had not had the opportunity to scan her brain, Jacqueline would have died on my watch. I will never, ever forget that phone call.

Looking back, can you think of an example where your intuition was warning you of something but you ignored it? When was the last time you had a really good hunch and went with it? On balance, how often do you trust your judgment versus asking for advice from others or feeling very confused and conflicted?

You can listen to your body and trust your gut even when it doesn't concern your own health. Both Jacqueline's friend and I had an instinct that something was very wrong. If you have a child, this will resonate with you. If you don't, then you may have felt it for yourself. If not, just know that you have this power within you. Nurture it and believe it. It's incredible.

Chapter 9

Motivation: Stay Resilient to Achieve Your Goals

"He who has a why to live for can bear almost any how."

Nietzsche

The basic human drives are sleep/wake, hunger/thirst and reproduction. On top of these, we have the things that particularly motivate us—this could be helping other people, intellectual challenge, financial success or innovating. There are also the more negative motivations, such as fear, revenge and anger, as well as addictions. We will be very aware of some of these motivators, but less conscious of those that are more insidious, such as a fear of abandonment or the drive to be perfect.

While motivation is what keeps us going when it would be easier to give up, resilience matches this drive with the ability to bounce back from adverse situations and adapt to cope better in the future. Having a clear "why" inspires us to look at obstacles flexibly when they stand in the path of our hopes. Defeatism isn't an option for a motivated person, so if you're serious about maximizing The Source and building a resilient brain it's important to understand your own motivations.

Purpose: The X-Factor

Okinawa in Japan is known for its residents' longevity. As a result, scientists have studied their lifestyle to find the clues to explain their long, healthy lives. What they discovered was that Okinawans have a strong sense of what they call *ikigai*—translated literally, it means "the reason I get up in the morning." A sense of purpose. A "why."[1]

Having a strong sense of purpose correlates with well-being; it ensures we are goal-directed, motivated by a desire for a particular outcome that gives us the tenacity to keep on going. This is a complex unconscious brain state that relates to survival and ensures we are not easily distracted or derailed by destructive habits or addictions. This is important, as anything from the ping of a text message all the way up to alcohol addiction or an eating disorder can derail us from our goals. The stronger our sense of purpose, the more the reward of moving towards that goal outweighs any possible distractions to our brain.

Having a strong underlying purpose also enables us to keep the bigger picture in mind when we're struggling with minor goals. This is a huge asset. People with a strong sense of purpose are more likely to be passionate. When I first heard Steve Jobs say: "If you follow your passion you will be successful," I felt it was easy advice for someone super-successful to offer. But when I made my career change, I experienced the simple truth of it for myself and now firmly believe it.

When aspiring young people ask me what they should study in college or which career they should choose, I feel that this is the best piece of advice I can give, because when things get tough it's our passion that keeps us motivated. Following our passion is the expression of a strong inner purpose. It's a huge mistake to pursue a certain amount of money or the material aspects of a lifestyle as our goal if this does not correlate with any underlying passion. I see this come to a head time and time again in people when some kind of crisis highlights that they never really loved what they were doing but needed to maintain a certain lifestyle. This is not a sustainable approach to life, as the lack of meaning and purpose eventually surfaces as an issue physically, mentally or emotionally and can lead to burnout.

Many of the most highly motivated and resilient people I have encountered have overcome childhood trauma in their past. Whenever I feel overwhelmed myself, I look to famous examples of people who have overcome adversity, such as Nelson Mandela and holocaust survivor Viktor Frankl, to help remind me how small my own problems are. If I'm feeling directionless or as if I can't see the light at the end of the tunnel, I always feel better after reading some of Mandela's writing or speeches. Here is one of his quotes about releasing negative emotions in order to let the good ones flourish:

> As I walked out the door toward the gate that would lead to my freedom, I knew if I didn't leave my bitterness and hatred behind, I'd still be in prison.

In my opinion, the connection between overcoming trauma and building resilience is no coincidence. Learning, early on, how to survive existential challenges—bereavement, divorce, being uprooted and moving away from friends and family—can lead to a strong internalized determination to thrive in spite of life's unexpected challenges.

Our failures, or "not yets," reframed in the context of an abundant mindset and the power of neuroplasticity really do make us stronger if we take ownership of them and determine to use them as grist for the mill—feeding The Source, rather than draining it. It's important not to let the hackneyed nature of this truism diminish its power.

To get to the bottom of our underlying motivators, we first need to be totally honest with ourselves. What do we really want out of life and why do we really want it? Plant this question like a seed in your brain and leave it there to germinate. A lot of psychological work will go on in the background that will help formulate your action board in Chapter 13. If you find this difficult to do then working with a coach or attending a workshop or retreat could help.

The Ups and Downs of Motivation

I know from personal experience what it feels like to lose motivation—this is what happened to me towards the end of my medical career. With dwindling motivation and a dawning realization that I needed to change career, I felt worn out, depleted and as if I could no longer be myself in my job. It crept up on me. At first, I had put it down to the standard doctor's lot: after all, almost everybody around me was exhausted and overworked, too. But gradually, my intuition told me there was more to it.

Interoception enabled me to read the physical signs that my body was running on empty: I felt constantly tired, resentful of colleagues that didn't pull their weight, and lacking intellectual stimulation. Also, I just couldn't visualize myself as a psychiatrist, some 10 or 20 years into the future. I couldn't see it, and it wasn't the future I wanted for myself. Aspects of myself that are important—creativity, mystery and autonomy; the very things that had inspired me to study psychiatry in the first place—felt neglected.

When I finally made the decision to retrain as a coach, I knew it was risky (and some of my friends and family viewed it as baffling and reckless) but I felt so motivated by my new goal that I was sure I'd made the right decision. I'd never felt so motivated about anything. Despite the considerable self-doubt I experienced on the coaching course (surrounded by a cohort of experienced business people, and feeling very much like a fish out of water), I was determined to succeed but it was hard. I cried the whole way through the mid-term assessment with my course tutor Jane, because I had so much emotionally invested in my career change and I desperately wanted it to work out. I was sure she would confirm that I was never going to make it and would be better off returning to medicine. Throughout, Jane listened and empathized then said, "Tara, I'm trying to tell you that you're doing brilliantly, but I don't think you can hear me." Later, once I had begun to build up my practice, she would tell me, "There's a certain level of drive and hunger that sets some people apart from the crowd and you had it."

People had often said I would make a good doctor, so I had heard this kind of thing before, but this time I chose to accept it, learn from it and harness it rather than bat away the compliment with some form of mock humility. I chose to cling to this rather than the negative internal voice that was trying its best to sabotage me. I was very determined not to waver and I had a strong sense of internal drive to succeed. I was reminded of a psychological theory on the potentially positive impact of adversity—an initial feeling of inferiority that makes you work extra hard and pans out to a high level of competence. I was ready to accept with humility that there was something bigger at play and that if I tapped into it I could create my ideal future.

For me, my hunger and drive came from a place where positive and negative emotions were aligned. I was determined to become a successful coach because I loved the work, and passionately believed my expertise meant I had something valuable to offer. But I was also motivated to change because I wanted to get away from being a hospital doctor. Other examples of positive and negative motivators aligning might be if you choose to leave a relationship because you feel sad in it and restricted by it, but also because you want to build a happier future and find somebody else who has the same values and aspirations as you. Similarly, taking a sabbatical, renting out your flat and going traveling may be escapist (you hate your job and are in denial about your relationship), but it's also motivated by a desire to have a life-changing adventure that will give you new perspective. A pre-getaway health kick is another example where motivation is a mixture of positive and negative (dreading being seen in your swimwear and wanting to feel confident).

However, some negative motivators just need to be nipped in the bud. In a day-to-day sense, surfacing what drags us down and acknowledging when we feel drained by someone or an activity can help us stay goal-directed and avoid destructive habits.

Beware negative motivators

Strong survival emotions—such as fear, disgust or shame—can often act as powerful motivators themselves. The brain's strongest drive is for our survival, and this is another way in which we haven't evolved appropriately from how we functioned when we lived in caves to what we need to do to operate effectively in the modern world. It's easy to mistake the influence of emotions such as shame or sadness, as the brain will "dress it up" as proactive choice. We may convince ourselves to stay in a bad marriage, for example, because we think it's the "right thing to do" for a host of reasons. In reality, though, the true motivation may be more fear- or shame-based: we simply don't want to be alone. The same is true for a career we have outgrown, or friendships that no longer make us feel good.

Negative motivators are most likely to creep in when we have a lull in momentum—when the going gets tough and it feels as if we're not getting anywhere with our goals. They may be acting at a subconscious level to sabotage our positive efforts. Knowing yourself, trusting your gut, developing mastery of emotions and making good decisions will help you to spot negative motivators at work and challenge them to create a better future.

LEE: THE CHALLENGE OF MINIMIZING DISTRACTION

Lee, a film-maker in her late twenties, came to me complaining of feeling overwhelmed. She had hit a wall when a project fell through. She knew she needed to pick herself up and uncover the motivation she had to make good work, but the setback had knocked her. She just couldn't focus. I asked her to create a list of distractions and motivation-sappers, as well as a list of what really motivated her. After a week of journaling, her list of de-motivators read like this:

- Checking social media and comparing myself to other film-makers.
- Scrolling mindlessly through my own social media feed to remind myself of happier times.
- Getting sidetracked with decorating and domestic projects when I'm working from home.
- Wasting time browsing dating profiles.
- Supporting a close friend who I love but who drags me down with her.
- Drinking too much in the evenings.
- Watching a lot of mindless TV.

In contrast, her list of motivators read:

- Watching an inspiring new movie.

- Contacting an old mentor to ask for advice and meeting her for coffee.
- Meditating.
- Working on a new idea for a whole morning, with all my tech notifications switched off.
- Going to an exhibition.
- Going for a run.

By raising her awareness of her distraction tactics and reminding herself of what keeps her on track, Lee decided to make some changes. For a while she had to keep referring to her list when she felt her motivation slipping but eventually she naturally turned to her motivators and put some rules into place around her de-motivators. She says it's still a work in progress but gets easier each time she has to make the effort.

Put aside two pages of your journal to create similar lists for yourself and see if there are any changes you can make right away.

Social media can either be inspiring or distracting. I'm a fan of it, but I think it's important to set boundaries around it. If you're tempted to check your smartphone mindlessly during the day, make this harder by deleting apps from your phone but keeping them on your tablet which you can peruse in the evenings. Numerous studies show that spending too much time online is bad for our mental health.[2] Do yourself a favor and set some controls around your usage. Anything that distracts you from your goals needs to be managed.

Perspective: Motivation's Silent Partner

Keeping a sense of perspective and taking a long view of life is helpful during stressful times, particularly when the gap between where you are and where you want to be feels insurmountable. Everyone is tested by challenging experiences, such as bereavement, heartbreak or financial difficulties; this is all part of the rhythm of life. It's sound advice to keep our problems in perspective and remember that there will nearly always be other people who are worse off than us. Yes, there will be those who are much better off as well, but we can still remember that ours aren't the worst problems in the world—even if it feels like it sometimes! If, for example, you can afford to buy this book; if you have access to books; if you have the time to focus on personal development, then you are already better off than most people on this planet. As touched upon previously, throughout history people have endured situations that many of us don't even have to begin to imagine today—slavery, apartheid or the Holocaust—so we can consider ourselves fortunate most of the time . . . if we choose to.

One question I often ask myself is, "How much will this matter in five years' time?" Perspective is about relativity in time as well as compared to the experiences of others. The answer— even when it concerns a seemingly major issue—is usually "not much" or "not at all." Another way to get a sense of perspective is to ask yourself what advice you would give your sibling or your younger self if they were in your situation. The act of making the issue less personal, yet related to someone you care for, changes our perspective in a way that makes the issue more manageable for the brain to process. This means that it is more likely to lead to better decision-making, because depersonalizing a problem makes it less threatening for our brains.

On the other hand, when the people I work with feel guilty about what we call "first-world problems," especially given that there are so many people living on and below the poverty line in this world, I reassure them that, just as our brain doesn't differentiate in its emotional response to an imagined scenario or a real one, all of our problems are real and challenging to us. And I heed my own advice, which means when I'm feeling like I should be Superwoman and soldier on regardless, I sometimes remind myself that if a patient were to come to me with x health concerns, y issues at home and z work stress, I would explain to them that there is a limit to what one person can endure! Perspective allows us to be a little kinder to ourselves, especially when equipped with the sort of well-developed coping strategies and the proactive approach that I have outlined in this book.

Having a sense of perspective helps us to grasp the fact that a mistake is only a mistake if we don't learn from it, and therefore train our brain not to repeat that pattern. When we feel like life keeps knocking us around, it helps to visualize ourselves as a snake repeatedly shedding its skin. We may have to go through this process over and over again, but we come out shinier and renewed each time we recover and learn from a difficult period.

Time for Action

It doesn't matter what your goal or intention is—it could be to live a balanced life or have better health or change your career—but strengthening your motivation will help you switch from imagination to action. If you want your dreams to become reality, you need to start doing something about it, and you need the resilience to keep on going with patience and the ability to move beyond distraction until you're there. How aware are you of what your goals actually are, what you need to do to progress towards achieving them, and what your possible barriers to this might be?

Developing an absolute belief in the abundance of possibility (remember Principle 1, page 25) and our potential power to realize our goals will help deepen motivation. When there is a potential abundance of money, love, success and fulfillment in our mind, we don't act from a "lack" mindset, which limits us. This is absolutely key to motivation, which is why I'm reminding you of it here. Thinking from this point of view helps expand our sense of what's possible. In order to manifest abundance, it's crucial to make space for it in our life. Sometimes this means making a leap—leaving a job or a relationship. At other times, it's subtler, requiring lots of smaller changes to facilitate huge transformation.

Chapter 10

Logic: Make Good Decisions

"The greatest mistake a man can make is to be afraid of making one."

Elbert Hubbard

We used to believe logical thinking is intrinsic: that we either have a "good logical brain" or we don't. Perhaps our parents and teachers decided for us where we sat on this spectrum as children and we have been wedded to this narrative in our head ever since—for better or for worse.

Broadly, we tend to think of the scientifically and mathematically inclined as more logical and analytical, and the artistic as being less so. Within modern society, being logical and analytical is massively overrated and being creative, intuitive or empathic far less so. The latter are even known as "soft" skills; but everywhere from the psychiatry clinic, through normal life to the most successful cohort of society, people find these skills much harder to master. Perhaps because they are more sophisticated and complex, or because we are taught to think logically from an early age. Either way, if we want to thrive in our life—especially given the rise of artificial intelligence and machine learning—it would be wise to focus most of our energy on trusting our gut, mastering our emotions and feeling in control of creating our own future.

The very fact that you are reading this book takes it as a given that you are good enough at being logical, but there may be an additional benefit to understanding how your logic can be derailed by strong emotions or unconscious biases.

The Left-Brain, Right-Brain Myth

For a very long time the neuro-myth of the brain's left–right functionalization has been the basis of many personality tests, self-help books and team-building exercises, but the science has moved on in terms of how we understand the brain to be organized. To unlock the mind and override some of these blocks, we need to understand the brain as a series of systems rather than a set of locations or a story of two halves.

We used to believe that logical, analytical thinking occurred in the left half of the brain, and creative or emotional thoughts in the right. However, modern neuroscience tells us that any sophisticated decision-making utilizes both sides of the brain and is integrative in its nature. Brain scans of people making decisions show many different and apparently unrelated parts of the brain firing at once when they are thinking about a complex problem. All information flows from left to right, back to front, bottom to top and, in all cases, vice versa. The more agile and healthy the brain, the better this whole-brain connectivity works.

It is simply not the case that everything to do with being analytical resides in one side of the brain, and everything to do with being creative in the other. It is also not the case that creative people use their right brain more and that those who are more logical use their left hemisphere more, or that left-handed people are more creative. A study using functional MRI (fMRI) scans (which show oxygen levels that correlate with activity in different parts of the brain) of 1,000 brains of people aged 7 to 29 by researchers at the University of Utah in 2013, found that it is the connections among brain regions on both sides that enable us to engage in both creative and analytical thinking.[1] Both sides of the brain tended to be equal in their activation, with roughly equivalent neural networks and connectivity.

This ability to look at healthy brains is a relatively recent development. We used to have to glean information about how the brain works from experiments on people with brain diseases. Experiments in the 1960s on people who had their corpus callosum (the bridge that connects the two halves of the brain) cut as a treatment for schizophrenia led to scientists determining which sides of the brain were largely responsible for language, arithmetic or art in their patients. While there is some truth in this, the advent of fMRIs of healthy brains has proven that the brain is a dynamic collection of systems and networks, cross-brain connections and complex lateral firing.

The Dangers of Logic

When our brain seeks to apply logic to a situation, it is attempting to tap into what it knows to be the "rules" of cause and effect: the idea that every action has a consequence. The positive side of this is that this should be about taking responsibility for our actions, being forgiving and learning from our mistakes—all of which are healthy states for our brain to be in. The negative side of this is that it can make us risk-averse and overcautious.

Logical decisions are perceived to stand in contrast to reckless, ill-thought-out ones, but this can make us think of logic as the enemy of risk-taking. In fact, logic should underpin and encourage a healthy appetite for taking considered risks, helping us to hone the ability to identify risks that are worth taking, moving us forward in ways that are bolder than the more obvious choices, and delivering growth without compromising stability and security. When weighing up the pros and cons of a big decision, it's important to try to remember this, although it's easy to get paralyzed by overthinking and decision fatigue. As a general rule, once we have made a decision, it tends to feel as if it was not as bad as we thought it would be. The key is to make a decision and then make it work.

The weighing up of logical decision-making is complex and sophisticated. It's also highly energy-intensive. It is perhaps surprising that although the rumination that leads up to a decision requires mental energy, it's the point of decision itself that is most energy-intense for our brains. This explains why reducing the number of unnecessary choices in our day (what to wear, eat, watch, react to on social media) is an effective way to conserve decision-making energy for bigger and more important decisions. This is known as "choice reduction" and can include a regular morning routine or laying out your outfit the night before, to avoid using up brain power on too many small things.

It's also important to note that a disposition towards logical thinking doesn't necessarily apply consistently across all areas of life. Think about the people you know at work and in your personal life. Can you think of examples of people who make great decisions at work, but whose personal lives have been blighted by bad decisions: chasing destructive friendships, mismanaging stressful family relationships, alienating their children? Many, many people excel at work but make a mess of their personal lives. This demonstrates that strong logic and emotional mastery do not necessarily go hand in hand. You have to work on each pathway to maximize The Source: there are no shortcuts, and you can't use strength in one pathway to compensate for the absence of another.

Pattern Recognition and the Brain

Let's look at what happens in our brain when we make a logical decision. In an ideal world, the decision will align in our brain physically, mentally, emotionally and spiritually, using all of our appropriate pathways equally. In reality, this is rarely the case, so we experience a discord in our neural pathways and have to rank them in order of significance, using its filtering process to reduce the potential negative consequences of risk-taking.

When faced with a what-to-do problem, our brain's pattern recognition system kicks in. This is a complex process that synthesizes information from the different parts of the brain, calling up memories of previous, similar decisions. We work through each choice rationally in our cortex, as the brain attempts to devise the best response now, measuring the present situation against previous ones. We ask the question: "Does this make sense based on the data I have available?" We then assess how this sits in our limbic system: "Does this *feel* right?" Our logical brain calculates likely outcomes and potential consequences, running a series of "What if?" scenarios like a chess player planning future moves.

To guide the way, emotional value tagging acts as a fluorescent highlighter, telling us which bits of past (and present) information to pay attention to. All of this is informed by emotional responses to the memories laid down in our past—what happened, what the outcome was and how our reaction to it led to success or failure. Each of our memories will be infused by a memory of how we felt at the time, thus informing our logical appraisal of the present moment. Our intuition either backs up or conflicts with the logical and emotional answers. We then decide which pathway(s) are least risky to go with, hence the almost permanent lingering doubt that we could have made a "better" decision. The key here is that every decision we make, however logical, is always biased by emotion.

There's no way to avoid our brain seeking out these tags, and the emotional element of decision-making is key—so much so that research shows that when the parts of the brain responsible for making sense of emotion are damaged, decision-making is slow and incompetent, despite the fact that the capacity for objective analysis is still there.[2] This illustrates the fact that logic cannot function optimally in isolation—it works collaboratively with other modes of thought, especially emotion.

There is something we can do to keep our logic on track: keep a skeptical eye on our "pattern recognition" system, casting The Source as a watchman, one that can act as a constructive challenge to emotional tagging, and sense checking any inferences we may make based on past experience.

Sometimes pattern recognition works well, but at other times, it can backfire spectacularly. In their book *Think Again: Why Good Leaders Make Bad Decisions and How to Keep It From Happening to You*, three management academics analyzed 84 bad decisions by successful leaders.[3] They found that in most cases danger set in when the brains of the leaders leapt to conclusions when faced with situations that called to mind previous experiences. What seemed logical conclusions to them were in fact dangerously flawed assumptions.

Some, even highly astute, people lack the ability to think around a decision in a holistic way. This may be a consequence of a rigid attachment to their own perspective, a lack of emotional intelligence or a strong subconscious bias at play. This is also a problem that comes up a lot in long-term relationships, particularly divorced parents, where one or both feels aggrieved but still has to interact with the "offending" party because of the children. Where we're entrenched in a set way of behaving and thinking, we can feel unable to see the situation from another point of view. We may cling to our righteousness, or the role we have taken on in the relationship (whether that's "the provider," "the carer," "the grown-up" or whatever it happens to be). Logic goes out of the window in this situation and bias takes over. False logic takes hold. This is dangerous, because "you don't know what you don't know." It takes a major effort of will, and conscious work on emotional attunement and self-awareness, to reset the balance, but it is possible to do so.

Spotting false logic

So, if it's easy even for the cleverest among us to make such epic mistakes, taking biases for truths, what's the best way to spot fake "logic" and recognize the real deal? In my opinion, the most important type of critical thinking directs itself inwardly to assess the trustworthiness of our own thoughts. Consciously raising memories of similar situations in the past that might be influencing our assessment of what's happening now (perhaps at a subconscious level) is step one. Then we need to consciously challenge the chain of comparative thinking, asking ourselves:

- What's different about now?

- Is my interpretation of what happened in the previous situation accurate?

- Could I think about the present situation from another perspective?

We need to counterbalance our own thinking and challenge our assumptions.

The ability to think around a problem, flexibly and adaptably, is logic in action. It works in multiple dimensions, and is the antithesis of "A to B thinking." The more we think in this rounded way, the better we get at it. From a neuroplasticity point of view, we need to retrain ourselves to tap into logic and rationality just enough to maximize and calibrate the other pathways, whenever we are faced with a new situation.

Developing an awareness of how our logic and decision-making pathways are constantly being influenced by everything we are subject to—from the people around us to our home environment and new learning—reminds us of the power we have to balance our sense of logic. When it comes to making big decisions based largely on logic, where do you think your strengths and weaknesses lie? The act of reading this book, and of taking on board the ideas contained within it, will have an impact on your ability to employ logic differently to make good decisions, although the extent of any lasting change will be determined by your actions: whether you put the suggestions and exercises into practice.

Chapter 11

Creativity: Design Your Ideal Future

"If you hear a voice within you say, 'you cannot paint,' then by all means paint, and that voice will be silenced."

Vincent van Gogh

Creating the life we want requires vision, and not just the vision to imagine the reality we want, but also to spot the potential opportunities that are occurring around us all the time and that could help us make our way towards that future. In this sense, creativity is not the traditional view of being good at art or full of new ideas, but it is the ability to shape our own brain by what we expose it to, designing our own future through proactive choice. Examples of this are everywhere when you start to look: from celebrities who have reinvented themselves, redesigning their identities—such as Victoria and David Beckham, Miley Cyrus, Mark Wahlberg, Angelina Jolie, Rihanna, Kim Kardashian, Justin Timberlake or Arnold Schwarzenegger—all the way up to the ground-breaking icons that have changed the world, including Abraham Lincoln, Nelson Mandela, Gandhi, Marie Curie, Mother Teresa, Martin Luther King and Emmeline Pankhurst.

Creativity is freedom. It enables us to direct the full power of The Source to create the life we envisage for ourselves. It enables us to draw on our other pathways, and use them in unexpected ways, using the law of attraction and visualization to manifest our desires.

A creative brain is one that can put ideas to use in unexpected ways, using contrasting combinations of thoughts to foster new ones. This is the new (and at the same time, ancient) superpower of the human mind: to reinvent, imagine, improve and rethink. When we're thinking with our whole brain, and devote our full creative power to a situation or problem, we see possibility where others see limitations. An Olympic sailing medalist I met at a party once told me: "When most people stand on a beach and look out to the sea at the horizon, they see the end. I see the beginning." Creativity gives us the power of interpretation.

To be creative we also need to develop a certain level of confidence in our right to express our unique take on things; to value our own ideas and interpretations. When I explain this to people it often strikes them as a pretty radical suggestion. I'm often met with the response: "But I'm not creative." This is frustrating and saddening. We are taught to think of creativity as something very narrowly defined (quite often linked to a natural talent for art). It is assumed, like a "logical brain," that you either have a "creative" disposition or you don't. I was told at school that I wasn't creative because I wasn't good at drawing. It turns out there is a generation of people afflicted by this myth.

This may have prevented you from starting your own business or even dressing in a certain way. The illusion around great artists of exceptional talent doesn't help, but dig a little deeper and even the most successful artist is likely to have "made it" as a result of tenacity, resilience and self-belief; that and a knack for spotting opportunities as they arise. You can take inspiration from this to turn the conventional notion of creativity on its head. Rather than being limited to artistic or cultural talent, it is about the ability to create your future, to be fully present and in charge of your life.

The neuroscience of creativity

Neuroscience is currently exploring what it is that characterizes creative minds. Researchers at Harvard recently identified a pattern of brain connectivity that is associated with idea generation.[1] In their study, people were put in a brain scanner and asked to come up with novel uses for everyday objects, such as a sock, soap or chewing gum wrapper. Some people were crowded with everyday, unimaginative uses, and found it hard to filter these out, so were likely to answer with the obvious examples, such as covering feet, blowing bubbles or containing gum. Highly original thinkers, in contrast, showed strong connectivity between three networks of the brain (mind wandering, focused thinking and selective attention), when they were thinking and came up with out-of-the-box ideas, such as a water filtration system, a seal for envelopes and an antenna wire.

Stop negative filtering

Mind wandering, focused thinking and selective attention can all be strengthened with practice. Giving yourself the time and space to think without distraction can lead to new ideas and perspectives. This is the benefit of actively allowing your mind to wander. By consciously bringing our desires, hopes and dreams to the forefront of our mind, our brain will be able to home in more effectively on opportunities that will lead to the outcomes we want. This is focused thinking and can be achieved through action boards and visualization. So often we filter out thoughts that don't fit our immediate purpose and we may unconsciously police thinking that feels "off the rails" or wacky, but in freeing ourselves up to think a broader range of thoughts we enable action. This is why we need to sharpen up our selective attention and filtering skills. In Chapter 9 we talked about having a "why"; here it is more a case of thinking "why not?" Why not apply for that new job or go on that date your friends are suggesting? Why not take up that new hobby you have been putting off for years? That new project idea that pops into your head when you're trying to focus on something else: write it down and return to it. Or you could try flipping a problem on its head to give it a new perspective. Try to catch yourself next time you're self-editing and ask yourself whether the thought you're pushing away could be a fruitful one. Remind yourself to think from a perspective of abundance rather than lack.

There is a practice called "rapid prototyping" which suggests generating as many ideas as possible, relegating the ones that don't work to the "try again in the future or in a different scenario" list and not stopping until we come up with the one that works perfectly for now. When I was considering starting up my own business, my ex-husband's uncle—a lovely man and serial entrepreneur—suggested starting a list of possible businesses and said that when it got to 100, there would be one option on the list that I could viably pursue. It took me two years to get to 100 ideas, but then I knew in my gut that coaching was the reason to leave medicine. I felt nervous and excited at the same time but mostly I felt confident that I would put everything into making this work to create the future I wanted.

Ban red-pen thinking

One of my most creative friends is an entrepreneur who has built a brand from nothing. She is always innovating and thinking up new schemes. Her partner is the same. When I asked her what her secret was she said simply: "There's no such thing as a bad idea in our house." She explained that she and her partner give each other and their kids permission to explore their ideas without shutting them down. Good ideas will naturally rise up and stand out. Stamping on them (yours or anybody else's) before you've had a chance to weigh them up can only be damaging.

Once we give ourselves permission to open up and play with lots of potential ideas and possibilities, creativity rewards us, enabling us to spot opportunities in unlikely places. It means we can sense when to take a chance and when to question or pursue something. It helps us hone a strong intuition, giving us the flexibility to recognize possibilities that might bypass us otherwise.

The following quote is often attributed to Kurt Vonnegut:

We have to continually be jumping off cliffs and developing our wings on the way down.

This illustrates the essential nature of creativity; it isn't a frilly extra—it is the resourcefulness to think your way through difficult situations and challenges and come out flying. And who doesn't want to excel at that?

Remember, you are creative already!

If you're still feeling a little skeptical about your creative abilities, look around you. You've created your home. You've created your career. You may have created a relationship and maybe even some children. You create your meals, topics of conversation, a welcoming atmosphere when guests come to visit, your garden, your friendships . . . the list goes on, and that's before we've gotten to any of the more obviously creative hobbies you may have. And if you don't think you have enough examples, try something new! Remember what we learned about the contribution of novelty to neuroplasticity (see page 104). What are your pre-conceived notions of how creative you are and in what way? We are all innately creative, and it's time to push that intrinsic power to be bold in the search for radical self-expression—the life you really want. Set a big picture challenge and watch your vision unfold. The exercises in the following chapters have been designed to allow you to unleash your creativity and use it to fuel your vision for the future.

PART 4

Fire Up The Source

"Whatever you do, or dream you can do, begin it. Boldness has genius, power and magic in it."

Goethe

And now for the exciting part! You've worked through the ideas in this book and understand how your brain pathways have evolved and how you can harness the power of neuroplasticity to strengthen them; you've learned about the transformative power of adopting an abundant attitude and understand how visualizing your ideal future could help you make it happen. Now it's time to put all these ideas into practice to realize your true intention.

Whether your aims are work-related, romantic or have more to do with general self-development, I'll be encouraging you to start disrupting your established pathways and creating change, using a mixture of simple and more involved exercises to turbocharge The Source.

These exercises will build through a four-step plan. You can follow this over four weeks or four months—you decide the pace that works for you. **The golden rule is only to move on to the next step once you feel you've received meaningful benefits and insights from the previous one.** Where appropriate, continue the actions of the previous steps in unison as you move forward.

The four-step plan is based around cognitive science, specifically the principle that lasting behavior change happens in four stages:

- **Step 1: Raised Awareness** (making the subconscious conscious and turning off your autopilot). You'll have thought a lot about this already as you've worked through the book. Hopefully, you are already feeling motivated to change. The exercises in Chapter 12 will further inspire you to expand your self-awareness, pinpointing the things about your behavior and thinking that are most in need of transformation.

- This provides the raw material to feed into **Step 2** (Chapter 13): creating the powerful **Action Board** to design your vision for the future and set your goals for change.

- Turning your imagined future into reality requires action. This is where **Step 3: Focused Attention** (Chapter 14) comes in. Practicing new behaviors and training yourself to think in new ways will be helped by you being more present, using mindfulness and visualization to help you devote your energy to the things that matter.

- **Deliberate Practice** (repetition) is the crucial final **Step 4** (Chapter 15), as you work on different aspects of The Source and embed new brain-friendly habits to ensure you can flourish, manifesting the full potential of The Source at its abundant best.

As you overwrite deeply embedded belief systems, your new ways of thinking will come to redefine the new you. Once this happens, you'll be better equipped to deal with the surprises life can throw at you, and you'll start to get more of what you want out of life.

I know from experience that when I work through these steps with people, the benefits quickly become cumulative. It's incredibly motivating once you start to feel and see the effects of the changes you are making, and this motivation will help you to stay on track. As you begin to live from this perspective of abundance and manifestation, your belief in your own power to make and maintain positive change will grow. You'll watch, in awe and excitement, as The Source directs your destiny.

Chapter 12

Step 1: Raised Awareness—
Switch Off Your Autopilot

"Until you make the unconscious conscious, it will direct your life and you will call it fate."

Carl Jung

love Carl Jung's quote on the previous page—it speaks to the heart of The Source and its potential to unlock our bright future. We are now ready to embark on the exercises and visualizations that will power up The Source and prepare to move our subconscious into our conscious. Make sure you clear some time to give yourself peace and quiet, free from distractions, so you can fully focus on yourself.

As we saw in Chapter 2, there is surprisingly little difference to the brain between an event actually being experienced in the outside world and a strongly imagined vision of the same event—whether it's something major or an everyday event. The exercises in this section will help you to build strong visualizations that engrain an imagined version of the future life you wish for deep inside your brain. In this way, The Source experiences the abundant reality you desire in advance of it happening, and so becomes more attuned to seizing opportunities, taking positive risks and making it happen in the future. Keep your journal at hand to record your insights from the exercises ahead.

Exploring Your Relationships and "Imprints"

The "models" we have for what family, love and "self" mean to us are set early on by our primary attachments. But their biggest influence comes as a result of the way we internalize these attachments, experiences and beliefs, then "imprint" them onto other relationships and situations as we move through life. This conditioning explains how a trigger and a response have become connected in our brain over time. The more we experience something, the more we lay down pathways for that connection in our brains through neuroplasticity and increased synapses. The brain's pattern recognition system then kicks in whenever it recognizes a new situation or relationship as "similar" in some way to one we have faced before.

This could include anything from our response to food, violence or perceived criticism, and each of us has a different response to these things because of what we have experienced growing up. Many people who were brought up in households where money was tight find it extremely uncomfortable to see food wasted; others often leave a meal unfinished and never eat leftovers. Some people stay in an abusive relationship because that is what they were used to. Some of us can take feedback and try to make it constructive, while others totally shut down at the slightest critical comment. Are the templates in your brain conducive to abundant thinking? Or might some of them be limiting and self-defeating?

Understanding this imprinting is helpful for understanding the way that the "past you" carries over into the present and dictates the future if we let it. It helps explain how any current relationship—at work, with friends or in intimate relationships—can trigger a response habituated in an earlier (childhood) setting.

Know your ghosts, change your destiny

Do you feel ready to begin exploring the ways your family and early attachments have shaped your neural pathways and how you view yourself, and how they have provided you with a set of expectations you project onto new people and situations as you come across them? Understanding your "ghosts" and the way they impact on the functioning of The Source is an important first step to shaking them off if they are holding you back.

Turn to a fresh page in your journal and start by filling in what the words below meant within your family and close relationships when you were growing up. What examples can you think of?

- Roles: What was your "role" in your family? What other "roles" were there and how did you relate to them? Examples might include "go-between," "scapegoat," "peacemaker," "rebel" or "deputy mother."

- Secrets: What were the secrets and lies in your family when you were growing up? Who kept them? How did they influence your life growing up? An example might be: "No one talked about Uncle Ray's drinking problem."

- Beliefs: What were the overriding beliefs in your family growing up? Were there any unspoken or unquestionable rules? Was there disagreement and conflict around differing opinions? People often mention concepts such as "hard work always pays" or "what goes around comes around."

- Values: What were held up as core "values" in your family? Were honesty, hard work, kindness, success, fairness, self-expression or intellectualism considered more important than anything else? Did this resonate with you?

- Boundaries: What was your family's attitude to boundaries: rules, illegal behavior, promise making and breaking, transgressions of all varieties?

Exploring the "ghosts" we all carry, and thinking about how well they serve you now, is a revealing and rewarding process. Are there ghosts you have taken on without questioning their helpfulness or accuracy? Have you found yourself following "rules" that conflict with your deep wishes? Note down any insights in your journal. Try to keep these insights in the front of your mind and notice when they play out in real life. Make journal notes on how they show up in the present. Start to make small changes that can adjust these sub-conscious reactions over time. This is how you start to take charge of your future.

CHLOE: CARRYING TOO MUCH

I did this exercise with Chloe, who is a mum of three in her thirties. In her family, growing up she was the "peacemaker" and she found herself in the same situation in her family life, mediating arguments between her children, as well as her husband and his brother.

She was exhausted emotionally when we began working together. She had reached a crisis point where she had realized she couldn't carry on supporting everyone else as she had been until now. She had to bring this role playing to the surface in her own mind and then consciously reset her boundaries.

It was one thing starting to unpick and change her own behavior, but the resistance to change she experienced from her family came as a shock. Disagreements intensified initially when she stopped intervening, and Chloe's children's attention-seeking behavior got worse as they tried to force her back into her old role. "They are making stuff up to get each other into trouble," she told me. "Not just occasionally, but all the time. They are exaggerating their reactions and the eldest has told me that I no longer care about her." Chloe sat her children down and explained that it didn't help them to come running to her when they were upset. In the real world, they would have to sort out their own disagreements. Eventually, the children began to complain less to Chloe and sort things out among themselves more. The household found a new level of calm.

She was both horrified at the mental games her children would play to keep her in the role they were accustomed to but also by the realization that she was creating a future of unhealthy relationships for them. Her love for them helped her to be tough in the short term and she went on to study a form of psychotherapy to further help her family, but also to counsel others to develop healthy boundaries.

With something as complex and interconnected as the networks in your brain and the way this influences your behavior, knowing where and how to start taking control can feel overwhelming. Looking into your own past conditioning is a good place to begin understanding the patterns that have been laid down in your brain to make sense of the world and your place in it. Now we are going to look at how these shape more specific themes and constructs that control and limit your life now, in order to ensure that they don't impact on your future.

Sense-check your self-limits

You've already looked at factors in your childhood and family that will have influenced negative thought patterns you hold, and the next stage is to explore our limiting beliefs further, and particularly their influence and relevance in your life today.

1. Turn to a fresh page in your journal and divide the page into three columns. In the first column, write a list of up to six limiting beliefs you hold about yourself. They are likely to be the stories you repeat to yourself or other people who are close to you, to explain what you feel are shameful or unhelpful actions or responses. Examples might include: "I'm not the creative type" or "I find it difficult to meet people." If you find this difficult, try using the sentence starters "I'm not . . ." or "I can't . . .".

2. Now take each sentence in turn and ask yourself the following: What is my evidence for believing this? Write these "supporting statements" in the second column next to the original self-limit.

3. In the third column, write down the evidence against this statement. Is it "true" in an objective sense? Use your past experiences to explore this statement fully. Consider what proof you have, if any, and how you could look at such "proof" in a more skeptical way.

4. Now ask yourself how helpful these beliefs are for your happiness.

Consider what these beliefs add to your life, and how they impact on your behavior and happiness. Do you want to hold these beliefs about yourself? If you don't, could you let go of them? What would help you to begin this process? Think back to your intention/big-picture goal (page 24) and visualize how changing these beliefs is likely to affect you achieving your big-picture goal.

After this exercise, make sure you give yourself a well-earned boost and take some time out to do something you enjoy, just for yourself. It's difficult to take such deeply felt criticisms and explore them with an open mind. Could you identify the friends and family who make you feel good and plan to see them soon?

This might be a good time to start a list of things you love about yourself in your journal. These might include "I love my independence; my creativity; my kindness; my vulnerability." Read your list of things to love regularly and remind yourself that you have disrupted the self-limiting beliefs whenever they pop into your head in the future. Put this technique of being able to challenge your own thinking into your toolkit until the time when it comes naturally to you—and it will.

As you move forward through the steps, if you feel as if you are slipping into a "lack" mindset, or beginning to tune into negative self-talk, turn back to your list and draw comfort and confidence from it.

Reframing Failure

Much more often than we realize, our perceived failures or vulnerabilities turn out to be some of the most important predictors of transformation and success in our lives, but at the time we are much harsher judges of ourselves than we would be of others. From an exam failure or redundancy that led to a successful career shift, to a relationship ending when it was time to move on, we are too quick to judge the unexpected nature of these actions as an indicator that we failed, rather than as part of the ongoing process of moving onwards and upwards.

Nurture The Source at these times by going back to the basic steps and focus only on what you can learn from any "mistakes." The only real mistakes, of course, are the ones we don't learn from.

Play around with alternatives you haven't considered before. Make sure you start in a low-risk environment that will not compromise work or relationships. As soon as it looks like a new approach is not going to work for now, shelve it and move on to the next idea—it may come in handy later. Ideas that are not an option now may work in the future.

The idea of failing fast and often is advocated by many entrepreneurs and successful technology companies such as Netflix and Facebook. Find ways to include an idea in your list of options that is like the joker in a pack. This could be as simple as trying a new hairstyle or glasses instead of your usual look; spending time with creative people if you are in a very corporate environment (or vice versa); or suggesting you go to an immersive art show or a walk with a friend instead of doing the usual coffee.

Create a list of achievements

In your journal, make a list of everything you have ever aspired to. This could include roles like wife or mother / husband or father; qualities like "having a voice" or expressing yourself creatively; or assets like wealth or success in your field. Underline those you have achieved. Look at the words and feel yourself fill up with a sense of achievement for the aspirations you have realized so far.

There may be some things on your list that you have been aspiring to for so long that you haven't stopped to acknowledge the fact you already have them. Own them: for example, "I am a loved and loving step-parent" or "I have built up a successful, stable business." Alternatively, ensure that you put down achievements that weren't on your to-do list, but that nevertheless demonstrate your endurance, capability, skills and determination.

Start a gratitude list

Turn to a double-page spread in your journal and write "MY ABUNDANT LIFE" in small but clear capitals across the middle of both pages. Over the course of the next few months, as your new habits bed in, fill this page up with all the things you have to feel grateful for. Fostering an attitude of gratitude primes The Source for abundance. The law of attraction is marshaled along by believing you already have the things you wish for. Growing this gratitude list in your journal is a great way to prime your brain to notice when good things happen. Add to it whenever you can—you could even make that part of your daily journaling task.

Moment to moment, rather than skipping over a flash of recognition or achievement, or a fleeting pleasure in a quest for "What next?," stop to feel gratitude for other people, circumstances or serendipity as well as your own qualities. This will harness your brain's value-tagging system and make positive achievements and happy thoughts easier to recall in the future. Doing this regularly will attune you to abundance.

Write in Your Journal

To get the maximum benefit from the four steps, you'll need to write in your journal daily about your thoughts and reactions to events and the people in your life. You don't need to write long entries, but aim to be honest and open about your emotions, motivations and behaviors.

Before bed tonight, spend a couple of minutes writing down some thoughts about your day in your journal. Then, note three positive actions you have made each day to help nurture The Source and create your ideal future. These can be as small as you like—from strengthening your emotional intelligence by thinking about a problem from another person's perspective to going for a mindful walk or reaching for a novel instead of your phone after dinner.

You could also begin to note down things that make you feel either energized or distracted and depleted. Start to explore your thinking around some alternative strategies to your most common stumbling blocks, e.g., "The next time a relationship makes me feel bad about myself, or the next time I make a mistake at work, I will do x instead of what I have done in the past." You could also think about the micro "failures" that occurred in your day: things you left unsaid; un-kindnesses you allowed; distractions that have taken over. What might you do differently next time? Shake up your autopilot and question your default settings. Channel the ideal version of how you'd like things to go tomorrow.

Each week in your journal, you should also choose three aims for the week. Choose a relationship (romantic or platonic) aim, a work aim and a personal development aim. These should be achievable small steps on the way to your big-picture goals. You'll already have strong ideas about what these big-picture goals should be and the rest of this book will help you develop them further. Examples of micro-challenges to aim for weekly might be:

- Relationships: Make an effort to listen actively more often to colleagues/partner. (Big-picture goal: Develop my emotional intelligence and empathy to strengthen key relationships.)

- Work: Speak up more about my ideas, or research potential mentors. (Big-picture goal: Start my own business.)

- Personal development: Commit to repeating a new affirmation each day to boost my self-esteem. (Big-picture goal: Stop self-criticism and be confident and happy in my life choices.)

Once you feel the benefits of these small changes, you'll be more inclined to push yourself further in the future—to take a bolder stance on your home, work and commute. Ask yourself big questions: Should I move to a new house, or stop renting and buy my own place? Should I be planning to quit my job, work from home or move closer to work?

As you move through the remaining steps and begin to visualize your ideal future, your answers to some of these questions will fall into place. Your journal will help you chart this journey, and explore the different possibilities for your life. Use it.

RAISED AWARENESS CHECKLIST

You have:

- Completed the "Know your ghosts, change your destiny" and "Sense-check your self-limits" exercises (pages 193 and 196) and filled in your journal with the results.
- Created a list of achievements and begun your gratitude list.
- Written in your journal daily and identified your three aims for the week.

You should be feeling full of positive energy and focused on all the things to like about yourself at this stage. This is a wonderful springboard into working on your action board, which is the next step in the process.

Chapter 13

Step 2: Action Board It

"Who at best knows in the end the triumph of high achievement, and who at worst, if he fails, at least fails whilst daring greatly, so that his place shall never be with those cold and timid souls who neither know victory nor defeat."

Theodore Roosevelt

Now it's time to create your action board. I'd like you to spend up to a week creating your board as it's a process that benefits from repeated bouts of attention. It's important to take your time to create a board that feels authentic to you, is inspiring and represents an accurate reflection of your innermost wishes. It's not something to rush, or to put together quickly with random positive images that you like—it must speak more deeply to you in terms of what the images represent, now and into the future.

What Is an Action Board?

An action board is a collage that represents everything you aspire to. Often called a dream or vision board, I prefer the term "action board" as we are looking to create something that will inspire and manifest in your future through your actions, rather than merely a vehicle for daydreams of second homes abroad and lots of money. We need to combine the positive desire with the energy and action of emotional intensity.

Creating an action board is about identifying your innermost dreams and representing them pictorially. However, as

I mentioned, this does not mean that you then just sit back and wait for things to happen—for the money to pour in, your ideal partner to sweep you off your feet, or your body shape or confidence to radically change as if by magic. You create an action board to prime your brain to grasp any opportunities that will bring you closer to the things you have identified you want in your life. More than this, you will also use the board to take action to make those dreams a reality. For example, if weight loss or achieving a particular body shape features on your board, an image relating to this will act as a prompt to encourage you to start going to the gym, do yoga or change your diet.

Those items that appear to be less within your control—such as getting married, pregnant or being spotted for a promotion—will fall into place once you start ticking off the first few things on your to-do list that are more immediately actionable. I like to think of it as like a pension plan where the more you put in, the more your employer puts in to match your investment. You should use it actively, not passively: your visualized future should inspire action now!

I have mentioned throughout the book that this is an important part of the process of directing the full power of The Source, and mobilizing your mind to envisage and create it. On the following pages we will go through the process of creating your board, with advice on how to choose the images that will make the most effective "symbols" for the life you want to build and how to use your completed board for maximum efficacy. Now is the time to start.

Throughout the book, I have invited you to start collecting images of your own to use. That's because action boards are incredibly powerful tools—images track instantly to your brain's visual centers, bypassing conscious thought, which means the brain's filtering system can't edit them out or dismiss them. They are emotive and symbolic, inspiring energy and action in the real world. Compared to a traditional written list of "personal goals" or even a "to-do" list, an action board will have way more impact on your brain and your future behavior. It might feel strange or even a bit silly to make an action board at first, but visualization and creating action boards are skills that you will feel more in tune with over time, as repetition builds and strengthens the pathways in your brain for these activities.

In this chapter, I am going to show you how to create an action board of your own, which will represent your hopes and dreams for the next 12–18 months, and will help prime The Source to set to work on making it happen for you. I can promise this experience will be life-changing but I will also say that at times it can be slow, frustrating and seemingly lacking in direction. When it feels like a struggle, return to the principle of patience we mentioned in Chapters 1 and 9. Your ability to deal with these ups and downs will say as much about you as what you put on your board.

My action boards

Action boards form a key part of my own practice, and I hope you will find them equally beneficial. It took me seven years to get to an action board that seemed to be just perfect for my life and keep me motivated (with a few updates), so patience is key. I had small wins along the way, but there was one action board that I made in 2015 for the new year of 2016 that really defined and celebrated a turning point in my life and indeed led to me writing this book.

Seven years ago, I was in the very early days of setting up my freelance practice, so it made sense to include an actual target figure of money that I wanted to earn on my board. Kate, a fellow coach who I often worked with in the north of England, staying in economical hotels and taking the off-peak trains back and forth, encouraged me to choose a number higher than just what I needed or felt was achievable. She suggested close to double my original figure. I thought she was a little too optimistic and I probably wouldn't be able to match the number that we agreed upon, but it would be great. Sure enough, the following year, her number was exactly how much I earned.

One year I chose a full-page accessories advertisement that featured a beautiful, glossy horse kicking up some water: I wanted my business to be a stable, strong entity that disrupted conventional thinking. That year I went from being freelance to setting up as a limited company; from working alone to having a team; and later becoming a professor at MIT Sloan School of Management, an award-winning author and the world's first neuroscientist in residence at Corinthia Hotel, London. I was also in sufficient demand with my thought leadership that I was frequently traveling around the world—now in style—to speak at conferences. These last few were things I could not have dreamt of when I picked that horse!

Finally, I realized that I had thrown myself into work as an avoidance strategy for the emotional instability caused by my divorce. In 2014, I had put a tiny heart onto my otherwise business- and travel-focused action board. This was all wonderful on the surface, but deep down I still wasn't ready for love and was very much living in the past.

By late 2015 I had revolutionized my thinking through a combination of soul-searching, a yoga retreat, long digital detoxes, total decluttering of negative people from my life, and following my own learnings on abundant thinking and visualization. In December, I started a totally fresh action board for 2016 with a strong sense of purpose. Previously I had sometimes simply added to the one from the year before, but this time I threw the old one out. I put an engagement ring in the top left corner and a phrase from a magazine advertisement (rarely but sometimes I use words if they really resonate with me) in the center, which read: "Joy comes out of the blue."

In February 2016, I met my now-husband on a plane flying from Johannesburg to London. I am sure that meeting in the sky counts as "out of the blue" . . . And he proposed nine months later. We were both people that had resolved never to marry again and indeed had not done so for nine and seventeen years, respectively. He claims (to anyone who will listen!) that he found true love for the first time in the autumn of his life. As I witness and share in his unreserved joy, I know there is hope for anyone at any age to achieve what they have always truly, deeply wanted.

How to Get Started

An action board will be the ultimate manifestation of priming your brain to design your life. The fact that you create it with your own hands and see it every day in full color activates numerous pathways in your brain (tactile, visual, emotional, intuitive and motivational), sending them the core message about what you truly want far more powerfully than just reading a list or thinking about your goals from time to time will ever do. Harnessing the concept of selective attention (see page 35), combined with the neuroplasticity of behavior change, reaps results.

You can create an action board on anything from a piece of letter-size paper up to a massive poster-size board. As well as the card, all you need is a stack of magazines or sources for images, scissors and glue or spray mount. You can look online for images but there is greater power in selecting images with the power of touch, so to speak. Even when you think you've found all the images you need, return to the magazines or get some new ones. This process benefits from not being rushed, stepping back and coaching yourself to fine-tune it. To create a really powerful impact on the visual centers of the brain, I suggest you use only images and possibly numbers—but try to avoid words. (Obviously, if you find a phrase or a quote that really resonates with you, you should use it.) One exception is the exact amount of money that you would like to earn; if this is important to you, that figure should appear boldly in numbers on your action board.

It is a good idea to use at least some metaphorical representations of what you want to achieve, rather than only direct representations or concrete examples. For example, if you are looking to move then attractive interior décor images make sense, but to trigger the emotional and subconscious parts of your brain as well as the logical and conscious parts, you could use less literal images. You could use a picture of a balloon to remind you of the importance of being free from burdens, or a chosen icon that speaks to you to remind you of the qualities of your best self.

Such metaphorical images are very powerful as they message your subconscious, allowing you to use abstraction and value tagging (see page 38) to spot and grasp opportunities that you might have missed otherwise. Just as you sometimes dream in symbols, with your subconscious creating metaphors to make sense of your experiences and thoughts, so you can direct your subconscious with imagery, particularly when it's non-literal. Metaphorical images also make your action board more private, and less obvious, which may make you feel more comfortable about placing it in a prominent position in your home.

Let your gut take the lead

To start making your action board, take the stack of images you have collected and group them on the floor or your desk according to their themes. Next, use your intuition to start placing some of these images onto the card—just loosely, don't glue them yet. Put the things that are most important to you in the center and/or near the top of your board. Group images in areas of the board according to whether they relate to work, love, health or travel, etc. You can keep the different life areas physically separate on the board or ensure that they are all touching and connected somehow if that feels right. Think about whether you want more space in your life and make sure your board is not crammed full if you do.

Once you have completed your first draft of the board, look at it as a whole. Step away and take some time out before coming back to it. Then remove any images from the board that don't feel right to you, even if they were particular images that you were drawn to when you first found them. Now go through all the magazines again and see if there are any images that you didn't notice the first time but which you are now drawn to. Find a place for them on your board. Once you have completed your second draft, leave it in a safe place (breeze-proof, pet-proof and child-proof!) for at least 24 hours.

A day later, or at the next opportunity you have, survey your board and make any final changes, then prepare to make it all stick! You might like to show it to one other person you trust before you immortalize it, asking them to challenge you about a few of the things on it, asking questions such as, "Do you *really* want this?," "Have you asked for as much as you deserve?" or "Is there anything else you want that you might have missed?" Once you are sure you have answered all these questions, paste all the images on to the board and, next, find the perfect place for it.

Where to place your action board

Your action board needs to be placed somewhere highly visible to you so that you see it at least once every day. The best places include by your bed so you see it every night before going to sleep, or inside your wardrobe door so you see it every time you get dressed in the morning. (This is a good option if you live in a shared house or simply do not wish everyone else to see it.) If you feel proud of it, it's better to display it and have nothing to hide; but this is not always possible or preferable.

A particularly good reason to place your action board by your bed is so that you can look at it just before you fall asleep at night. The transitional period from wakefulness to sleep is known as the hypnagogic state of consciousness. Mental phenomena such as lucid thoughts or lucid dreaming occur during this "threshold consciousness," making you particularly suggestible at this time. If you have paid deliberate attention to a repetitive activity just before falling asleep—particularly something that is new to you—then this will tend to dominate the imagery of your dreams. This is known as the "Tetris effect" (named after the 1980s video game, this occurs when people devote so much time and attention to an activity that it begins to pattern their thoughts, mental images and dreams) and is yet another way to imprint on your subconscious the things that you should be looking out for during the day.

Take advantage of the fact that an element of novelty creates a more powerful effect on your brain and, before you turn out your light and go to sleep, look at your action board intently, using a mantra or affirmation (see page 238), or simply state out loud what the metaphorical images represent to you. Do this frequently for the first month you use the board. After that, just let its images sink into your subconscious with a glance.

If you wish to store it privately yet be able to see it regularly, one idea is to take a photograph of your action board with your smartphone and make it your screensaver. An online way to make and store an action board is by creating one on an online platform such as Pinterest and then saving it as a private board that you can look at on your phone or tablet frequently. However, you will have to make extra effort to ensure that you look at it often enough for it to work its magic on your brain's pathways.

When to create an action board

The best time to start an action board is, of course, right now! Your birthday or New Year are other obvious times to start it, or the start of a new project/phase of your life or the academic year. If you've never made one before—or not made one for a while—then start now and use it until the end of this year or even the end of next year; then choose a regular time to update it or make a new one approximately annually.

Your action board does not necessarily have to remain static for the whole year, although personally, I find that it takes up to a year for most things to become a reality—as your brain pathways strengthen, directing your behavior to create that new reality for you. In the meantime, you will need to achieve a balance between patience and determination.

Action Boards in Practice

Once you've created your action board and used it to the extent that its images have really sunk into your psyche, it will start to act like a visual directory that accompanies the lists in your journal—of the things you most want to achieve and the daily notes on the three steps you've taken every week to help you get there. Each time you review your board and see something else on the board that is now a reality in your life, add that feature to your accomplishment list. Watch your accomplishments list grow and use this as a guide as to whether you need to reinforce, remove or add something to your board, depending on how your year is panning out.

ACTION BOARD CHECKLIST

You will have:

- Created a powerful action board filled with images that accurately represent the life you dream of.
- Looked at it every day, and ideally multiple times a day, to refresh your memory and allow the images to work on The Source.
- Visualized the items on your board coming true.

Chapter 14

Step 3: Focused Attention— Neuroplasticity in Action

"The act of paying attention contains tremendous power."

Deepak Chopra

Now that you are starting to pinpoint the behavior and thinking patterns that are holding back The Source, you're ready to begin to train yourself to think in new ways. One of the simplest ways of clearing old patterns in order to create new neural pathways is to be more in the moment and more capable of focused attention. But this is easier said than done. This will form your goal for this step, as presence is something you need to practice regularly, and it can be done in a range of ways.

We will also delve further into the concept of abundant living and how you can firmly place yourself on that path to a brighter and happier future.

What Is Presence?

Put simply, becoming more present is the process of bringing our attention to experiences occurring in the current moment, which can be developed through meditation, mindfulness or other forms of training.

I prefer to think of presence as a way of life rather than discrete practices. Personally, I consider mindful eating, mindful walking and paying full attention to people as we interact with them to be as important as formal mindfulness practices such as yoga or meditation. This everyday presence often gets forgotten and I encourage people to make this the focus of their mindfulness practice, rather than obsessing about getting traditional meditation "right." Having said that, a daily practice, even for a few minutes, will make a radical difference to your brain. By the end of this week, you'll have created a personal mindfulness practice you can return to every day to nourish The Source. You'll soon begin to feel the difference.

My journey to presence

I started to become interested in mindfulness meditation in my mid-thirties when I had become increasingly disillusioned with work and life and felt a lack of direction and focus. My attention was easily distracted by the next new and interesting diversion and deep down I knew I was drifting away from who I really was and what I wanted to stand for. I knew about mindfulness from my parents growing up—there was a prayer room in our house where my parents would go to pray and meditate, lighting incense and sitting in silence, or doing breathing exercises. This was a daily practice for them but I never tried it and regarded it as just another one of the culturally diverse things my parents did. But in my late twenties, I started to recall this more as my close friends and I became interested in yoga. Sometimes in a yoga class, the teacher would talk about mindfulness, and I'd read the odd interview with a celebrity who swore by it, but I stuck to the yoga with a short meditation at the end.

However, as the evidence from brain scanning reached the point where I was starting to speak about mindfulness meditation at banks and hedge funds, I felt that I should have sufficient practical experience of it myself to be able to advise others.

Despite having practiced yoga regularly for ten years, I had to use an app and earphones for nine months before I could just sit on the Tube and guide myself through a 12-minute meditation. Now I run guided meditations in both my classes at MIT Sloan School of Management and in other workshops with businesses.

Individually, I have times when I meditate for 12 minutes most days of the week (mostly on the Tube), and times when I have not practiced formal meditation for weeks (although I always try to eat mindfully and draw myself back to my breath whenever I need to). Over the years I have found it easier to pick up the meditation when I am very tired, jet-lagged or under pressure and that the practice I do when I have more time gets me through the stressful periods. I consider it a way of future-proofing my brain when I can, rather than feeling guilty when I don't make time.

When I feel stressed about not finding the time, I remind myself of a wonderful story of a monk advising an executive to meditate for an hour a day. When the reply is that at busy times this won't be possible, the monk simply states: "At those times you must do two hours a day." The irony is not lost on me.

The Science of Presence

Once you start to work on becoming more "present," you'll feel calmer surprisingly quickly. And within two to three months of increasing your presence and starting a meditation practice, you'll have fundamentally changed your brain—that is the incredible effect of a regular practice.

From a neuroscience point of view, the evidence is clear. Clinical studies have shown both physical and mental benefits of mindfulness in healthy populations as well as patients with illnesses such as depression, anxiety, stress, addiction and post-traumatic stress disorder. Scans on people who use mindfulness meditation show significant neuroplastic changes in their brains.[1]

A consistent meditation practice increases folding in the brain—and its surface area. These changes are situated in the cerebral cortex, the layer of the brain responsible for processing and regulating data from the outside world. Committing to devoting a few minutes each day to meditation will give you a new clarity of perspective on what and who are your real priorities in life, supporting your "higher level" brain regulation and improving your resilience, making you more considered and balanced in your approach. If you want to maximize the power of The Source, mindfulness really is a no-brainer.

When I'm working with type-A business people who are very dismissive of anything they consider "fluffy," I always quote a study on the US Marines, which showed that those who practiced mindfulness meditation for 30 minutes per day had increased resilience after stressful combat training compared to those who did not.

A follow-up study looked at 320 Marines who were preparing for deployment to Afghanistan.[2] Half the group were given an eight-week mindfulness course, including homework and training on interoception—the ability to accurately "read" the body's signals that we explored in Chapter 7. They were encouraged to develop better awareness of bodily sensations, such as a churning stomach, fast heart rate and tingling of the skin.

Part of the training involved simulated combat experiences with live-action scenarios in a false Afghan village, where actors played Afghans and real-life conflict situations were played out. During and after this, a team of researchers monitored the blood pressure, heart rate and breathing of the mindfulness-trained and non-mindfulness-trained Marines, and took note of their neurochemical reactions to stress.

The mindfulness-trained group was calmer during and after the exercise, and reacted more quickly to a threat when it appeared. The Marines' brains were observed using MRI scans, revealing that the mindfulness-trained Marines had reduced stress-related activity patterns in regions of the brain responsible for integrating emotional reactivity, cognition and interoception. In other words, building the brain–body connection gives exponential benefits both physically and mentally. This is the crux of unlocking The Source.

Other studies show that meditating for as little as 12 minutes a day has a significant impact.[3] When people are resistant to introducing a mindfulness practice, I always say that I understand the impracticality of some of the more time-consuming well-being activities, but I honestly can't believe there is anyone that cannot find 12 minutes on most days to do something that will potentially change their lives. Most people don't disagree with that!

Find an app

I really wish for you to turn mindfulness from something that you have on your list of things you must do, to something that you want to find a place for in your life several times a week. One of the easiest ways to do this is to find a mindfulness app that appeals to you—then you'll have little excuse not to plug in and go with it at a time of day when you think you're most likely to be able to establish a regular time for yourself.

Try a few different mindfulness apps. There are plenty out there (Calm, Headspace and Buddhify are examples), but it takes a bit of shopping around to find one that works for you. If you are time-pressed, work out a "dead" pocket of time you can use—commuting is ideal. Don't get too hung up on the time you spend meditating as studies show that frequency, rather than duration, is the key factor, so ten minutes every day is likely to be more beneficial than longer but more intermittent sessions.[4]

LINDA: A SIMPLE INTERVENTION

Sometimes simply listening to a piece of calming music can have a similar effect to meditating, as it forces you to devote focused attention to one of your senses.

I once had a client, Linda, who was so stressed at the start of a coaching session that she was visibly agitated, talking at a hundred miles per hour. I knew I had to help her to change her emotional state in the moment, so I asked her to listen to a piece of classical music on my phone for three minutes with her eyes closed. At the end of just three minutes she sighed and said she felt much calmer, and she did come across as more present. We discussed that this was an easy thing she could do for herself between meetings as a task switch and mindful moment.

Linda also suggested she could bring her favorite mug into work and make an herbal tea to savor at the next meeting rather than grabbing coffee to go between meetings and further fueling her anxiety and indecision. Although this was a small, simple action, filling her mug every day grounded her and reminded her of home. It became an everyday act of self-care that reminded Linda to pause.

Developing a mindfulness practice allows us to quiet the noise around our brain, enabling us to press a pause button, calm our emotions and thoughts, and think more from a perspective of abundance. It's one thing to understand the benefits of a mindful life in the abstract, but you have to try it yourself to fully understand it.

BODY SCAN

While we're thinking about mindfulness and presence, we're going to return to the body scan exercise we practiced on page 136. There, I recommended trying to do this exercise every day for a week, and noting the results. If you haven't already tried this, start now. It'll enable you to be immediately more present in your body.

Did you notice any areas of tension? Did one side of your body feel more relaxed than the other? Were you able to let go of tension when you consciously tried to do so? Make notes in your journal on how you feel after the scan over the course of the week.

Over time, consider what signs your body is showing you that you need to take better care of yourself, such as aches or pains, low energy levels or skin problems. The more you practice this body scan, the more you'll find yourself, quite naturally, begin to tune into your body the rest of the time, too. By doing this exercise you'll begin to build a quiet but constant connection between your mind and body.

Delight in your senses

In addition to mindfulness, yoga at home or in a class for 30–90 minutes would be an excellent addition to a varied mindfulness regime, as it connects brain and body, but if this is something you need to work to find time for in your life, then choose two or three moments when you can just "be in the moment" fully and write them in your diary to make them an event.

Sensory stimulation is a great way to jolt your brain back towards mindful presence when you're feeling strung out or distracted. Easy and enjoyable ways to tune into your senses that might enrich your life and make you feel calmer and happier might include:

- Going for a mindful walk and looking at the color of the sky, leaves and flowers.

- Really savoring your cup of tea.

- Conjuring up a favorite childhood smell in your present home.

- Introducing new textures and patterns into your décor or clothing.

- Listening to evocative music or doing a spin or 5Rhythms class and getting back in touch with your body.

You will know what feels most appropriate for you and will fit into your life—perhaps it's a simple thing like buying a bunch of flowers each week for your desk at work to bring color, scent and a smile to your working day. Look for two or three ways to bring your senses to life during the week.

Novelty and sensory experiences, particularly those that are immersive and will free up your brain's mind-wandering mode, prompt the default network we explored on page 57.

Anything that encourages abstract thinking will increase the chances for The Source to seize opportunities that otherwise may pass you by when you're in the habit of overthinking and overanalyzing.

Live Your Intention

So far, we've concentrated on mindfulness as a powerful tool to introduce greater neuroplasticity to our daily lives, and it is really a game-changing experience once you get it. Now let us look back at your intention and aim, with an abundant mindset, and plan how to bring it to life.

Roadmap to abundance

This three-column exercise will result in actions that should form part of your weekly routine. You'll need to give yourself a forum for holding yourself accountable to making change happen for this exercise to work. This should be through your journal, an app or regular check-ins with a supportive friend or partner.

Start by brainstorming some ideal future outcomes based on your set intention and your insights so far. For example, you might decide that in your ideal future:

- I feel confident about life and don't let doubts hold me back.

- I am healthy, happy and in control.

- I get my dream job or start my own business.

- I meet my perfect partner and we start a family.

1. Find some large sheets of paper (e.g., flipchart paper) and colored marker pens. This works better on bigger paper than in your journal, but you can note down the outcomes in your journal afterwards. Decide what shape you would

like your ideal future to take and write this down at the top of the first page. This will tie in with your intention, and you might add more detail on other aspects of your life also. Example statements might include: "I'm happily settled in a committed and loving relationship," "I'm regularly called on to speak at events and I'm confident doing so" or "I've turned my hobby into a thriving business."

2. Now divide the rest of the page into three columns. The heading of the first column is BARRIERS. (Do not read ahead to the titles of the next two columns until you have completed this one.) In this column, list all possible barriers to your ideal state. Really push yourself to fill at least one page and if possible carry on to a second page. This is really important because unless you list all the possible barriers to your success, the exercise won't be as effective as it could be. Barriers could include things such as not having enough time, not earning enough money, being too shy, the very idea of it feeling like it's too much hassle, being too busy or lacking in motivation.

3. The heading of the second column is OPPOSITE STATE-MENT. Now list all the extreme opposite statements for everything listed in the first column—even if these statements could not possibly be true. For example, depending on what your first column holds, these might include phrases such as: "I have all the time in the world," "I have unlimited money," "I don't care what people think," "It would be fun and easy," "This activity is my absolute priority," etc. Be really bold and have fun with this column as you have nothing to lose—and it will inform the third column, providing you with more options for what your brain can do if unleashed from the barriers that are holding you back.

4. The heading for the third column is WHAT I DO DIFFERENTLY. Here you should list the real-life, day-to-day activities that would occur as a result of the statements in column two being true. These can include physical actions, thoughts or interactions with others. Examples might include: "I make time to go on three dates a week," "I pay to get my CV professionally updated," "I spend more time with my friends and have the confidence to ask them to introduce me to new people," "I create a website and really put myself out there" or "I invest in a new date outfit."

5. Finally, group the concepts in column three into themes. Now pick two or three actions to start with that you can implement from today, such as networking, clearing time to devote to your aim or your daily gratitude list. Tear off and scrunch up your barriers, and throw them away. Note down your insights in your journal.

Evidence-based Visualizing

The next exercise takes the principle of being the best you—or you on a good day—further and is based on research carried out at the National University of Singapore.[5] Contrary to what we once thought, not all visualization and meditation techniques produce the same effect on the mind and body. This study examined four types of meditation. These included two types of Vajrayana meditation taken from Tibetan Buddhism, namely a visualization of self-generation as a deity (imagining yourself with the powers of a god or goddess), and Rigpa, which literally means "knowledge of the ground" (grounding practices aim to attain Rigpa and integrate this into everyday life). The study also examined two types of Theravada meditation practices, these being Samatha (focusing on a single point) and Vipassana (gaining insight).

The researchers collected ECG (heart), EEG (brain) and cognitive testing data from participants in the study. From this information, they observed that Theravada and grounding practices produced enhanced parasympathetic activity in the autonomic nervous system, i.e., relaxation. However, Vajrayana visualizations did not lead to relaxation, but produced an immediate, dramatic increase in performance on cognitive tasks, as well as a more alert and active state in the body. This proved that the different categories of meditation are based on different neurophysiological responses.

Although there are massive benefits from practicing types of yoga and meditation that can decrease stress, release tension and promote deep relaxation or even sleep, I would strongly recommend that you try the visual meditation outlined below, which uses a similar technique to Vajrayana meditation, to bolster you in situations where you need your brain to perform at its best. I have found it very useful, as have many people I have worked with.

Research is ongoing into whether the radical enhancements in brain performance after just one session can lead to permanent effects with regular practice and which specific elements of the visualization lead to the cognitive benefits.

Identify with a powerful icon

The Vajrayana practices remain very sacred and secret to the Tibetan monks and can only be handed down through specific training and initiation. As this is a secular book, and out of respect for the fact that you are only supposed to visualize self-generation as a religious icon if you have been initiated in that religion, I would suggest instead that you pick someone that you know and respect, such as a historical character or a current secular figure. It does not matter if the person you choose is male or female, the same gender as you or not; it just has to be someone whom you view as being incredibly powerful in terms of having qualities that you would like or feel you need. Examples could include one of your grandparents, an author, activist or famous entrepreneur. Write this person down in your journal.

1. Find a quiet and secure place to sit or lie down with your eyes closed. Before you close your eyes, you may wish to look at a picture of your chosen icon for a moment.

2. Now, with your eyes shut, start to imagine this person in front of you. Imagine every tiny detail of what they look like and the presence they create in front of you.

3. Practice until you feel you could reach out and touch them, talk to them—or, if you opened your eyes, they would be there right in front of you.

4. Once you have mastered this stage, move on to imagining yourself as this person. Start by visualizing yourself from top to toe as your icon: their hair, voice, posture and mannerisms—and the power that you are longing for.

5. Feel this connection resonate throughout your entire body until you feel inseparable from the person. (This might take you many weeks to achieve.)

6. Keep practicing this visualization until you feel convinced that you now possess the qualities you were looking for, and can conjure them up when needed.

The purpose of this exercise is ultimately to realize through visualization that you already possess those very same qualities deep inside The Source. Directing your pure, focused attention in this visualization will encourage you to "fill up" with the positive energy you want to manifest.

Releasing burdens

Loss aversion (which defines "lack" thinking) is one of the strongest gearings of the brain and explains the brain's tendency to default to mistrust. Our survival emotions have a stronger effect on the brain than our attachment emotions, meaning that perceived losses have twice the effect on us psychologically as equivalent gains.

In order to manifest your ideal life, therefore, you need to train your brain to devalue any potential losses and filter out the unnecessary warning signs that your limbic brain may send to the front of your mind (the pre-frontal cortex). It's important to do this to ensure we don't get caught out by these red flags. The following visualization helps you do just that by releasing buried negative thoughts. You could read out and record this visualization on your phone and play it back to yourself or ask someone to read it out as a guided visualization for you.

Visualization: hot air balloon

Find a quiet, peaceful place where you will not be interrupted. Start by taking four deep breaths, then breathe normally, counting your breaths from one to twelve before performing the body scan exercise on page 136.

1. When you have completed the body scan from the tips of your toes to the crown of your head, imagine yourself

standing at the edge of a lavender field in Provence. Feel the warm breeze tickling your skin, hear birds singing in the distance and inhale the air pungent with the smell of lavender. Engage all your senses in your surroundings— the feel, smell, sounds, sights and even taste of the lavender field on this summer's day in the south of France.

2. As you look into the distance where the field of green and purple meets the huge blue sky, you notice a hot air balloon on a small hill. You walk over to it and soon you can see the weave of the basket and the rainbow colors of the silk balloon.

3. When you reach the hot air balloon, you realize that it is tethered to the ground by four sandbags fixed to the basket by rope. Notice the color of the rope and the size of the sandbags. Immerse yourself totally in this visualization.

4. A small rope ladder on the side of the hot air balloon's basket allows you to climb up into it. Once inside you see that it is very easy to steer and fly the balloon, but that you will have to release the sandbags if you wish to fly.

5. As you turn to untie the first sandbag you notice that it has FEAR written on it in large black letters. Untie the sandbag and watch it roll away down the small hill until it disappears into the distance. The basket hovers a little way off the ground.

6. Turn to the second sandbag and before you untie it notice that it has the word ENVY written on it in red letters. Untie the bag and watch it fall to the ground and burst, releasing all its contents, which evaporate into the atmosphere. The basket is now hovering clear of the ground but leaning to one side.

7. You go to untie the third sandbag and see that it has EXPECTATIONS written on it in purple letters. Untie this bag and hear it fall to the ground with a thud and disappear. Now you are flying some distance from the ground and there is only one thing left that is holding you back. I cannot tell you what is written on the fourth sandbag but it is the biggest personal burden that you are carrying. Read what it says [this will be unique to you]. Pause to take this in fully, then untie the final sandbag and watch it disappear into the distance as your hot air balloon starts to rise up into the sky.

8. Steer the balloon and fly as high or as low as you like. You choose where it goes. It could rise all the way up to outer space or float over oceans and mountains. You decide. Visualize your hot air balloon journey for approximately five minutes.

9. When you feel ready, slowly bring the hot air balloon in to land. This could be on a beach, back in the field where you started—anywhere you like. Step out of the balloon and focus your mind's eye back in your physical body. Count your breaths backwards from 12.

10. Breathe normally for a moment, then take four deep breaths and start to wiggle your fingers, followed by your toes. Slowly open your eyes and make a note in your journal of what the burden was that you released in your fourth sandbag.

We will not use an image of this on your action board but you could find an image of what release from that burden would look like for you—or simply find a picture of a hot air balloon to go on your action board, as I have for the last few years.

To me, the balloon represents freedom from burdens, from old, unhelpful patterns of behavior and from negative people.

Return to this visualization as often as you can, and particularly when any of the things written on your sandbags are weighing you down.

This exercise will help turbocharge the power of your visualizations, using the images on your action board to empower yourself to take proactive steps towards change unimpeded by obstacles that have held you back in the past. This is where focused attention meets action—and it's at this juncture that Step 4 of my process takes over.

FOCUSED ATTENTION CHECKLIST

You will have:

- Done your body scan meditation and continued to do this for a week to note the changes in your body and mind.
- Experimented with a range of mindfulness apps to find one that works for you and begun a regular practice.
- Completed the "Delight in your senses" and the "Roadmap to abundance" exercises (pages 222 and 224).
- Completed the "Identify with a powerful icon" and "Hot air balloon" visualizations (pages 227 and 229).

Chapter 15

Step 4: Deliberate Practice—
The Source Comes to Life

"There's no fate but what we make for ourselves."

The Terminator

In neuroscientific terms, the pathways in our brains are interconnected and multidimensional. It isn't helpful to think of neural pathways as linear A to B structures. They are influenced by a cocktail of deep behavioral conditioning, environmental influences, contagion and a range of other factors. My four-step process of raising awareness, envisaging change, focusing your attention and taking action with deliberate practice offers a 360-degree approach to change. Each step supports the others, and by committing fully to them all, you will transform the way you think and behave.

By raising your awareness in Step 1, you will have surfaced a lot of your subconscious autopilot behavior and made it possible to address it. Steps 2 and 3 will have helped you visualize the future you want and will support you to begin to attune your attention in a focused way.

In this final step, you'll be combining your new ability to focus with your awakened awareness of the life you want so that you can direct the full power of the force towards change. The exercises in Step 4 are all focused on deliberate practice, supporting you to turn thought and insight into action that will help you manifest the future you dream of. Regular visualizations will boost and sustain this.

ALEX: A PERSONALITY TRANSFORMATION THROUGH DELIBERATE PRACTICE

I was approached to work with Alex, a CEO in his forties, on the recommendation of his HR director. Despite carving out a successful career, many of his team found it difficult to work with him. Short on interpersonal skills and empathy, while being intensely demanding and micromanaging staff, led to years of rumblings turning into serious issues, with both the Chief Financial Officer and Chief Technology Officer threatening to leave. When I began working with him, Alex was unconvinced I could help.

"I've always been this way; I don't think I can change. It isn't my problem if people don't like working with me. I do a good job. Being a CEO means I need to be on top of everything, I need to worry for everyone before the worst can happen. I'm the bearer of bad news so it's no wonder people don't like me. Why should I let that bother me?"

I explained to him that the qualities that set great leaders apart are more often the less-valued skills of flexible thinking and emotional intelligence. I also explained that compassion and awareness of other people was certainly something we could train him to become more attuned to, and that it would just require some work. I also talked to him about the negative language he used and his controlling tendencies. I challenged his idea that being a strong leader had to be aligned to rigid control and constant emphasis on negatives.

Alex worked on his emotional intelligence, making efforts to give his team more positive feedback and encouragement and letting go of the control he had craved before by breaking his micromanagement habit. Alex had always worked hard on his relationship with all the people in his life—his children (two boys aged eight and ten), his wife, his assistant and his team. He was also good with customers, so I knew he could do it. He applied himself with all the effort he had put into his business. Where he had previously found some of his sarcastic comments to his assistant funny, he now realized they could be hurtful. And where he had once described a night in with his kids as "babysitting," he now felt the rewards of being more available to them as they embraced him into their hearts and minds in ways he hadn't even realized he was missing out on. They even started calling him before their mum sometimes when they needed help—something they would never have done before.

Alex began to share his positive vision for the future of the company and to celebrate the successes of those within his team. He made sure his senior managers knew he trusted them to do the job well, and he stepped back. As new people joined the team with no preconceived views of Alex they found him to be warm and even funny. Seeing this inspired the rest of the team to change their view of Alex. In response to their new-found openness towards him, he flourished, and repaired his relationship with the senior leadership team who had barely tolerated him previously.

Alex's story shows you must be prepared to disrupt yourself positively and stick at it to manifest deep change, to clear any remaining obstacles to change and to channel the abundant energy of The Source to power game-changing thought and action. It is absolutely possible to change fundamental things about yourself if you really want to and are prepared to put in the work.

Stamp out your abundance enemies

Once you have set out on the road to abundance, how can you ensure you stay on track? First, you'll need to identify things that you think really need to change. This is more about overwriting long-standing patterns than setting a new intention. Both are important to move forward with momentum. You may have noted some of these down in your journal already when you looked at your self-doubts and barriers to moving forward. They may be goals you have had for a while but repeatedly failed to achieve. Examples might include: get fit; find a new job; or get married.

1. In your journal, create a grid with three columns and three rows. In the first column, list each of your goals. In the second, honestly record any behaviors that are sabotaging your goal. In the final column, record the underlying belief that is motivating this counterproductive behavior. What might you believe deep down that supports your reluctance or inactivity? It could be a cynical feeling that you can't control a situation or the feeling that you don't have the energy to implement positive change now, or it could be a more insidious, deep-rooted belief that you aren't worthy of the future you desire.

2. Once you have filled in the columns, take a long hard look at the pattern you see before you. This has been created by a combination of your genetic inheritance, how you've been brought up and the choices that you have made—both actual choices in the real world and the level of mastery you have exhibited over your emotional and behavioral response to the outcomes. These have all shaped The Source.

3. The final column contains your abundance enemies. Take a look at them. What could you do to disrupt them? Could you create affirmations that overwrite your self-sabotaging beliefs with positive ones or change your behavior to carve out more time for activities that nurture your energy rather than deplete it? Be specific and, for each goal, list a specific action you can complete this week.

Create personal affirmations

Choose from the phrases in the third column of your "Roadmap to abundance" exercise (page 224)—ones that have inspired you before or compliments you have been given; any mantra that appeals to you now.

You can work with as many affirmations as you like, but I usually find three or four is plenty, and makes them easier to memorize and stick to. Here are some I have used myself:

- "I'm alright, right now" (from *The Power of Now* by Eckhart Tolle)

- "Things are exactly as they are meant to be" (from astrologer and author Lyn Birkbeck)

- "This too shall pass" (probably originating with the Persian Sufi poets but used by the English Poet Edward Fitzgerald and President Abraham Lincoln in his inaugural speech)

- "This isn't real" (from the *Divergent* movie scene in which Tris overcomes her fears)

- "No one can hurt me now" (from a friend who is an amazing female scientist and entrepreneur)

- "Things make sense later—sometimes much later" (from me)

As you can see, you can create personalized mantras out of phrases or quotes taken from books, films, conversations with friends or even your own insights on your journey of personal discovery, and use them to shift learnings from the non-conscious to the conscious part of your brain. Any new insights from your abundance exercises could be pivotal in unleashing The Source when used as a mantra.

1. Write your affirmations in your journal and type them into your phone so you have them with you throughout the day, or write them on Post-it notes and stick them by your bed, in the bathroom or kitchen: anywhere you will see them.

2. Aim to repeat them, mindfully, several times a day. Visualize them as true.

Push your boundaries

Working through the steps brings cumulative and major shifts of thought and behavior, but we can also take small steps to aid neuroplasticity and ensure that we are pushing ourselves out of that autopilot zone and embracing change in all areas of our life.

Fear of failure keeps us frozen and stuck and is the enemy of positive action. One powerful antidote is to begin to experiment more and push your own boundaries, moving yourself out of your comfort zone and off your autopilot more and more often so you get used to taking regular, healthy risks. There are really big examples of discovery through experimentation and "failure" which we touched on earlier. If you want to build up your experimental side, you don't need to start big.

Novel experiences increase neuroplasticity and, particularly when shared with a friend or partner, are a great mood-booster.

Try a new sport, go to a different park for your daily dog walk, change your standard radio station or pick up a book that you wouldn't normally read. Creativity and integrative thinking come about via the cross-brain connections you can make: the greater the diversity of your experience, and the wider your frame of reference, the richer The Source grows.

Tonight, try a new recipe for something you have never cooked before. If you have always stuck to recipes then cook *Iron Chef* style with whatever you find in your fridge and cupboards. If you are feeling bold then invite someone over to share the results of this experiment with you.

Every change that you embrace—big and small—helps The Source to gain confidence in the power of change and move away from that bias towards security and preserving the status quo.

Pause to consider your legacy

People like Mandela, Gandhi, Mother Teresa or Emmeline Pankhurst won't be remembered for having children (or not), helping their neighbors with their shopping or supporting family and friends in their lives, but for the overwhelming impact they have left on the history of humanity. But equally, raising kind human beings with integrity and purpose and supporting those around us will make the world a better place. We achieve this by nurturing the brains of our children, families and social or professional groups. Stop to consider what you do to make the world a better place, and how you could do more of it.

1. Visualize yourself as an old person towards the end of your life. Take the time to fully immerse yourself in this

visualization and imagine how you feel, what you are wearing and where you are sitting or standing.

2. Now, ask yourself how you feel. What are you proud of having achieved? What are the stand-out moments of your life? Who around you is truly important to you?

3. Make some notes about your responses and how they reflect on your current lifestyle and ideal future. Add a picture to your action board that relates to the passions and purposes you have uncovered, and explore ways that you can turn this into reality in your life.

Visualize The Source in you

This is an extra, final visualization to add to your others. Return to it whenever you like. It's a great way to envisage yourself with The Source at maximal power.

1. Start with the body scan (page 136) then take five deep breaths. Now breathe normally, counting your breaths in your mind from one to twelve—breathing in for one, out for two, and so on. With each breath visualize your feet walking down a stone step. Imagine yourself standing in front of a door roughly carved into a mountain wall. Open the door and step inside. As your eyes grow accustomed to the dim lighting, you realize that you are standing in a huge cavern with five full-length mirrors in it. Notice the color of the walls, whether or not there are windows around the room, and the shape of the mirrors—they might be rectangular, oval or another shape.

2. Walk up to the first mirror. In this mirror, your reflection is wearing your favorite sports jersey. There is something

about your poise and posture, the glow of your skin
and the tone in your muscles that tells you that you are
tension-free and at your peak of physical fitness. Examine
your reflection and drink in this image of strength,
stamina and inner calm.

3. Walk over to the second mirror. In this mirror, you are
   wearing only your underwear. The flatness of your belly,
   the light in your eyes, the glossiness of your hair and the
   plumpness of your skin tell you that you are healthy and
   well because you have taken time and care to nourish
   and hydrate your body. Send a mental photograph to
   yourself of what it is that you notice about yourself at your
   healthiest.

4. Walk over to the third mirror. In this mirror, you are
   wearing the perfect work outfit: this could be the
   sharpest business suit and shoes that you have ever
   owned, surgical scrubs, stylish smart-casual, whatever
   works for you. The ease of your stance and the scene
   in the background tell you that you have reached the
   pinnacle of career success, with all the comfort and
   security that go with it, whatever that looks like to you.
   You exude confidence from every cell of your body.
   Notice all the details in this reflected self, so that you can
   remember what this feels like.

5. Walk over to the fourth mirror. In this mirror, you
   are happy and relaxed and surrounded by the
   people whom you love and who love you. You are
   in your favorite social setting and wearing casual,
   comfortable clothes. The sound of laughter, the look
   of joy on your face and the obvious feeling of love

between the people make you feel warm and fuzzy. Bottle that feeling.

6. Walk over to the fifth mirror. In this mirror you are fit, healthy, confident, successful, happy and loved. A combination of all the attributes shown in the previous four mirrors. But this is not a mirror, it is a portal. Walk through the portal into your new life. Healthy, happy, confident, loved. Bask in this for as long as you like.

7. As you exit the portal you find yourself standing with your back to the stone door. You know that something has changed in your life forever. Something good.

8. Count your breaths backwards from 12—with each breath visualize your feet climbing back up the steps. Breathe normally for a moment, then take five deep breaths and start to wiggle your fingers, followed by your toes. Slowly open your eyes. Make a note in your journal of anything that struck you about this exercise, and cut out and note any images that you could add to your action board to represent these things.

And that is the end of the last stage, as we push our small and bigger steps towards a new life with deliberate practice. It has been a relatively short time since you began the practical aspect of *The Source*, but make sure that you take some time to yourself to look at what changes you have made so far, what awareness and enlightenment you have brought to your actions and behaviors—past and present. Read through your journal and review the insights you now have into your pathways and motivations, and the steps that you are taking to build your future.

DELIBERATE PRACTICE CHECKLIST

You will have:

- Identified your abundance enemies and devised three actions to vanquish them.
- Created personal affirmations to inspire and motivate you as you progress on your journey.
- Begun to try new things on a regular basis to stretch your comfort zone.
- Considered your legacy.
- Visualized The Source in you—you at maximal power.

Conclusion

Sustaining The Source

I've lost count of the number of people I've worked with who have reached out to me weeks, months and sometimes even years down the line to let me know that the action board they built based on this four-step process has materialized in the real world. They email me pictures of weddings, babies, product launches, promotion news, new homes: success and happiness in all its many, glorious forms. All of these are proof of The Source functioning on full: a thriving trinity of brain, body and spirit.

Keep visualizing and keep making it happen. Watch and marvel as your action board becomes reality, and the life you visualized begins to materialize, building a cumulative power that means that you can attract and achieve more and more each year.

Own and be proud of your ability to change and grow. People often tell me, "You changed my life." And I respond, "Thank you, but you did it." It is your awareness, your actions and your beliefs that will invoke change. Imagine what your life will look like in five, ten, twenty years' time as you strive to make your dreams reality, continuing to evolve in the way you have in the short weeks since you picked up this book. Enjoy it and believe it.

When you have completed all four steps consecutively, return to and read the final paragraph overleaf. Take a deep breath and let it out with a sigh. Feel the tension slip away from your muscles.

You've done it! You've created a new person—one whose life now has a different trajectory to the one you were set on when you picked up this book. You understand now that you can attract the things you want into your life. The world has so much more to offer. Your amazingly malleable, abundant and agile brain helps you spot opportunities and create and attract untold positive experiences into your life. You know you deserve this abundance, and you have no hesitation in grasping and making the most of exciting potential adventures. Where once you were stuck in entrenched patterns and belief systems, you've evolved to find a new freedom of thought. You've done this simply and with integrity. Integrity of your brain, body and spirit. You are The Source—the creator of your life.

Nothing can stop you now.

Notes

Introduction

1 Harari, Y.N., 2015. *Sapiens: A brief history of humankind.* Vintage.

Chapter 1

1 Kahneman, D. and Tversky, A., 1984. Choices, values, and frames. *American Psychologist, 39*(4), pp.341–50.

2 Simons, D.J. and Levin, D.T., 1998. Failure to detect changes to people during a real-world interaction. *Psychonomic Bulletin & Review, 5*(4), pp.644–9.

3 Ronaldson, A., Molloy, G.J., Wikman, A., Poole, L., Kaski, J.C. and Steptoe, A., 2015. Optimism and recovery after acute coronary syndrome: a clinical cohort study. *Psychosomatic Medicine, 77*(3), p.311.

4 Park, N., Park, M. and Peterson, C., 2010. When is the search for meaning related to life satisfaction? *Applied Psychology: Health and Well-Being, 2*(1), pp.1–13; Cotton Bronk, K., Hill, P.L., Lapsley, D.K., Talib, T.L. and Finch, H., 2009. Purpose, hope, and life satisfaction in three age groups. *The Journal of Positive Psychology, 4*(6), pp.500–10.

5 McDermott, R., Fowler, J.H. and Christakis, N.A., 2013. Breaking up is hard to do, unless everyone else is doing it too: Social network effects on divorce in a longitudinal sample. *Social Forces*, 92(2), pp.491–519.

6 Christakis, N.A. and Fowler, J.H., 2007. The spread of obesity in a large social network over 32 years. *New England Journal of Medicine*, 357(4), pp.370–9.

7 Sterley, T.L., Baimoukhametova, D., Füzesi, T., Zurek, A.A., Daviu, N., Rasiah, N.P., Rosenegger, D. and Bains, J.S., 2018. Social transmission and buffering of synaptic changes after stress. *Nature Neuroscience*, 21(3), pp.393–403.

Chapter 2

1 Clark, B.C., Mahato, N.K., Nakazawa, M., Law, T.D. and Thomas, J.S., 2014. The power of the mind: the cortex as a critical determinant of muscle strength/weakness. *Journal of Neurophysiology*, 112(12), pp.3219–26; Reiser, M., Büsch, D. and Munzert, J., 2011. Strength gains by motor imagery with different ratios of physical to mental practice. *Frontiers in Psychology*, 2, p.194.

2 Ranganathan, V.K., Siemionow, V., Liu, J.Z., Sahgal, V. and Yue, G.H., 2004. From mental power to muscle power—gaining strength by using the mind. *Neuropsychologia*, 42(7), pp.944–56.

Chapter 3

1 Gholipour, B., 2014. Babies' amazing brain growth revealed in new map. *Live Science*. www.livescience.com/47298-babies-amazing-brain-growth.html [accessed 24 September 2018].

2 Live Science Staff, 2010. Baby brain growth reflects human evolution. *Live Science*. www.livescience.com/8394-baby-brain-growth-reflects-human-evolution.html [accessed 24 September 2018].

3 Hirshkowitz, M., Whiton, K., Albert, S.M., Alessi, C., Bruni, O., DonCarlos, L., Hazen, N., Herman, J., Katz, E.S., Kheirandish-Gozal, L. and Neubauer, D.N., 2015. National Sleep Foundation's sleep time duration recommendations: methodology and results summary. *Sleep Health*, *1*(1), pp.40–3.

4 Thomas, R., 1999. Britons retarded by 39 winks. *The Guardian*. www.theguardian.com/uk/1999/mar/21/richardthomas.the observer1 [accessed 7 October 2018].

5 Lee, H., Xie, L., Yu, M., Kang, H., Feng, T., Deane, R., Logan, J., Nedergaard, M. and Benveniste, H., 2015. The effect of body posture on brain glymphatic transport. *Journal of Neuroscience*, *35* (31), pp.11034–11044.

6 Black, D.S., O'Reilly, G.A., Olmstead, R., Breen, E.C. and Irwin, M.R., 2015. Mindfulness meditation and improvement in sleep quality and daytime impairment among older adults with sleep disturbances: a randomized clinical trial. *JAMA Internal Medicine*, *175*(4), pp.494–501.

7 Danziger, S., Levav, J. and Avnaim-Pesso, L., 2011. Extraneous factors in judicial decisions. *Proceedings of the National Academy of Sciences*, *108*(17), pp.6889–92.

8 Watson, P., Whale, A., Mears, S.A., Reyner, L.A. and Maughan, R.J., 2015. Mild hypohydration increases the frequency of driver errors during a prolonged, monotonous driving task. *Physiology & Behavior*, *147*, pp.313–18.

9 Edmonds, C.J., Crombie, R. and Gardner, M.R., 2013. Subjective thirst moderates changes in speed of responding associated with water consumption. *Frontiers in Human Neuroscience*, *7*, p.363.

10 Begley, S., 2007. *Train Your Mind, Change Your Brain: How a new science reveals our extraordinary potential to transform ourselves.* Ballantine Books, p.66.

11 Alzheimer's Society, n.d. Physical exercise and dementia. www

.alzheimers.org.uk/about-dementia/risk-factors-and-prevention
/physical-exercise [accessed 7 October 2018].

12 Voss, M.W., Nagamatsu, L.S., Liu-Ambrose, T. and Kramer, A.F., 2011. Exercise, brain, and cognition across the life span. *Journal of Applied Physiology, 111(5)*, pp.1505–13.

13 Hwang, J., Brothers, R.M., Castelli, D.M., Glowacki, E.M., Chen, Y.T., Salinas, M.M., Kim, J., Jung, Y. and Calvert, H.G., 2016. Acute high-intensity exercise-induced cognitive enhancement and brain-derived neurotrophic factor in young, healthy adults. *Neuroscience Letters, 630*, pp.247–53.

14 Firth, J., Stubbs, B., Vancampfort, D., Schuch, F., Lagopoulos, J., Rosenbaum, S. and Ward, P.B., 2018. Effect of aerobic exercise on hippocampal volume in humans: a systematic review and meta-analysis. *Neuroimage, 166*, pp.230–8.

15 Rippon, A., 2016. What I've learned about the science of staying young. *Telegraph*. www.telegraph.co.uk/health-fitness/body/angela-rippon-what-ive-learned-about-the-science-of-staying-young [accessed 2 October 2018].

16 Abbott, J. and Stedman, J., 2005. Primary nitrogen dioxide emissions from road traffic: analysis of monitoring data. *AEA Technology, National Environmental Technology Centre. Report AEAT-1925.*

Chapter 4

1 Langer, E.J., 2009. *Counterclockwise: Mindful health and the power of possibility*. Ballantine Books. [See also: Alexander, C.N. and Langer, E.J., 1990. *Higher Stages of Human Development: Perspectives on adult growth*. Oxford University Press.]

2 Taub, E., Ellman, S.J. and Berman, A.J., 1966. Deafferentation in monkeys: effect on conditioned grasp response. *Science, 151*(3710), pp.593–4; Taub, E., Goldberg, I.A. and Taub, P., 1975. Deafferentation in monkeys: pointing at a target without visual

feedback. *Experimental Neurology*, 46(1), pp.178–86; Taub, E., Williams, M., Barro, G. and Steiner, S.S., 1978. Comparison of the performance of deafferented and intact monkeys on continuous and fixed ratio schedules of reinforcement. *Experimental Neurology*, 58(1), pp.1–13.

3 Gaser, C. and Schlaug, G., 2003. Brain structures differ between musicians and non-musicians. *Journal of Neuroscience*, 23(27), pp.9240–5.

4 Begley, S., 2007. *Train Your Mind, Change Your Brain: How a new science reveals our extraordinary potential to transform ourselves.* Ballantine Books.

5 Woollett, K. and Maguire, E.A., 2011. Acquiring "the Knowledge" of London's layout drives structural brain changes. *Current Biology*, 21(24), pp.2109–14.

6 Sorrells, S.F., Paredes, M.F., Cebrian-Silla, A., Sandoval, K., Qi, D., Kelley, K.W., James, D., Mayer, S., Chang, J., Auguste, K.I. and Chang, E.F., 2018. Human hippocampal neurogenesis drops sharply in children to undetectable levels in adults. *Nature*, 555(7696), pp.377–81.

7 Boyd, L. 2015. TEDx Vancouver, Rogers Arena [TEDx Talk].

Chapter 5

1 Siegel, D.J., 2011. *Mindsight: Transform your brain with the new science of kindness.* One World Publications.

Chapter 6

1 Goleman, D., 1996. *Emotional Intelligence: Why it can matter more than IQ.* Bloomsbury.

2 Killingsworth, M.A. and Gilbert, D.T., 2010. A wandering mind is an unhappy mind. *Science*, 330(6006), p.932.

3   McLean, K. 2012. The healing art of meditation. *Yale Scientific.* www.yalescientific.org/2012/05/the-healing-art-of-meditation [accessed 24 September 2018].

Chapter 7

1   Ainley, V., Tajadura-Jiménez, A., Fotopoulou, A. and Tsakiris, M., 2012. Looking into myself: Changes in interoceptive sensitivity during mirror self-observation. *Psychophysiology, 49*(11), pp.1672–6.

2   Farb, N., Daubenmier, J., Price, C.J., Gard, T., Kerr, C., Dunn, B.D., Klein, A.C., Paulus, M.P. and Mehling, W.E., 2015. Interoception, contemplative practice, and health. *Frontiers in Psychology, 6,* p.763.

3   Lumley, M.A., Cohen, J.L., Borszcz, G.S., Cano, A., Radcliffe, A.M., Porter, L.S., Schubiner, H. and Keefe, F.J., 2011. Pain and emotion: a biopsychosocial review of recent research. *Journal of Clinical Psychology, 67*(9), pp.942–68.

4   Hanley, A.W., Mehling, W.E. and Garland, E.L., 2017. Holding the body in mind: Interoceptive awareness, dispositional mindfulness and psychological well-being. *Journal of Psychosomatic Research, 99,* pp.13–20.

Chapter 8

1   Mayer, E.A., 2011. Gut feelings: the emerging biology of gut–brain communication. *Nature Reviews Neuroscience, 12*(8), pp.453–66.

2   Steenbergen, L., Sellaro, R., van Hemert, S., Bosch, J.A. and Colzato, L.S., 2015. A randomized controlled trial to test the effect of multispecies probiotics on cognitive reactivity to sad mood. *Brain, Behavior, and Immunity, 48,* pp.258–64.

3   Kau, A.L., Ahern, P.P., Griffin, N.W., Goodman, A.L. and Gordon, J.I., 2011. Human nutrition, the gut microbiome and the immune system. *Nature*, 474(7351), pp.327–36; Kelly, P., 2010. Nutrition, intestinal defence and the microbiome. *Proceedings of the Nutrition Society*, 69(2), pp.261–8; Shi, N., Li, N., Duan, X. and Niu, H., 2017. Interaction between the gut microbiome and mucosal immune system. *Military Medical Research*, 4(1), p.14; Thaiss, C.A., Zmora, N., Levy, M. and Elinav, E., 2016. The microbiome and innate immunity. *Nature*, 535(7610), pp.65–74; Wu, H.J. and Wu, E., 2012. The role of gut microbiota in immune homeostasis and autoimmunity. *Gut Microbes*, 3(1), pp.4–14.

4   Foster, J.A., Rinaman, L. and Cryan, J.F., 2017. Stress & the gut-brain axis: regulation by the microbiome. *Neurobiology of Stress*, 7, pp.124–136.

Chapter 9

1   Buettner, D., 2012. *The Blue Zones: 9 lessons for living longer from the people who've lived the longest*. National Geographic Books.

2   Dokoupil, T., 2012. Is the internet making us crazy? What the new research says. *Newsweek*. www.newsweek.com/internet-making -us-crazy-what-new-research-says-65593 [accessed 3 October 2018]; Twenge, J.M., Joiner, T.E., Rogers, M.L. and Martin, G.N., 2018. Increases in depressive symptoms, suicide-related outcomes, and suicide rates among US adolescents after 2010 and links to increased new media screen time. *Clinical Psychological Science*, 6(1), pp.3–17; Thomée, S., Dellve, L., Härenstam, A. and Hagberg, M., 2010. Perceived connections between information and communication technology use and mental symptoms among young adults—a qualitative study. *BMC Public Health*, 10(1), p.66.

Chapter 10

1 Nielsen, J.A., Zielinski, B.A., Ferguson, M.A., Lainhart, J.E. and Anderson, J.S., 2013. An evaluation of the left-brain vs. right-brain hypothesis with resting state functional connectivity magnetic resonance imaging. *PloS One*, *8*(8), p.e71275.

2 Bechara, A., Damasio, H. and Damasio, A.R., 2000. Emotion, decision making and the orbitofrontal cortex. *Cerebral Cortex*, *10*(3), pp.295–307.

3 Finkelstein, S., Whitehead, J. and Campbell, A., 2009. *Think Again: Why good leaders make bad decisions and how to keep it from happening to you.* Harvard Business Review Press.

Chapter 11

1 Beaty, R.E., Kenett, Y.N., Christensen, A.P., Rosenberg, M.D., Benedek, M., Chen, Q., Fink, A., Qiu, J., Kwapil, T.R., Kane, M.J. and Silvia, P.J., 2018. Robust prediction of individual creative ability from brain functional connectivity. *Proceedings of the National Academy of Sciences*, *115*(5), pp.1087–92.

Chapter 14

1 Gotink, R.A., Meijboom, R., Vernooij, M.W., Smits, M. and Hunink, M.M., 2016. 8-week mindfulness-based stress reduction induces brain changes similar to traditional long-term meditation practice—a systematic review. *Brain and Cognition*, *108*, pp.32–41.

2 Johnson, D.C., Thom, N.J., Stanley, E.A., Haase, L., Simmons, A.N., Shih, P.A.B., Thompson, W.K., Potterat, E.G., Minor, T.R. and Paulus, M.P., 2014. Modifying resilience mechanisms in at-risk individuals: a controlled study of mindfulness

training in Marines preparing for deployment. *American Journal of Psychiatry*, *171*(8), pp.844–53.

3 Hurley, D., 2014. Breathing in vs. spacing out. *New York Times Magazine*. www.nytimes.com/2014/01/19/magazine/breathing-in-vs-spacing-out.html?_r=0 [accessed 3 October 2018]; Wei, M., 2016. *Harvard Now and Zen: How mindfulness can change your brain and improve your health*. Harvard Health Publications; Rooks, J.D., Morrison, A.B., Goolsarran, M., Rogers, S.L. and Jha, A.P., 2017. "We are talking about practice": the influence of mindfulness vs. relaxation training on athletes' attention and well-being over high-demand intervals. *Journal of Cognitive Enhancement*, *1*(2), pp.141–53.

4 Basso, J.C., McHale, A., Ende, V., Oberlin, D.J. and Suzuki, W.A., 2019. Brief, daily meditation enhances attention, memory, mood, and emotional regulation in non-experienced meditators. *Behavioural Brain Research*, *356*, pp.208–20.

5 Amihai, I. and Kozhevnikov, M., 2014. Arousal vs. relaxation: a comparison of the neurophysiological and cognitive correlates of Vajrayana and Theravada meditative practices. *PloS One*, *9*(7), p.e102990.

Bibliography

Below is a list of books that have shaped my learning.

Begley, S., 2007. *Train Your Mind, Change Your Brain: How a new science reveals our extraordinary potential to transform ourselves.* Ballantine Books.

Coelho, P., 2012. *The Alchemist.* HarperCollins.

Doidge, N., 2008. *The Brain That Changes Itself: Stories of personal triumph from the frontiers of brain science.* Penguin.

Finkelstein, S., Whitehead, J. and Campbell, A., 2009. *Think Again: Why good leaders make bad decisions and how to keep it from happening to you.* Harvard Business Review Press.

Goleman, D., 1996. *Emotional Intelligence: Why it can matter more than IQ.* Bloomsbury.

Haanel, C.F., 2013. *The Master Key System.* Merchant Books.

Harari, Y.N., 2015. *Sapiens: A brief history of humankind.* Vintage.

Hesse, H., 2017. *Siddhartha.* CreateSpace.

Hill, N., 2004. *Think and Grow Rich.* Vermilion.

Ibarra, H., 2004. *Working Identity: Unconventional strategies for reinventing your career.* Harvard Business Review Press.

Johnson, S., 1999. *Who Moved My Cheese: An amazing way to deal with change in your work and in your life.* Vermilion.

Ramachandran, V.S., 2012. *The Tell-Tale Brain: Unlocking the mystery of human nature.* Windmill Books.

Sacks, O., 2011. *The Man Who Mistook His Wife for a Hat*. Picador.

de Saint-Exupéry, A., 2018. *The Little Prince*. Vintage.

Siegel, D.J., 2011. *Mindsight: Transform your brain with the new science of kindness*. One World Publications.

Tolle, E., 2001. *The Power of Now: A guide to spiritual enlightenment*. Yellow Kite.

Acknowledgments

I would like to thank Zoe McDonald for all her patience, understanding and brilliance in helping me to realize my story.

The team at Penguin Random House UK are outstanding and I would like to thank Joel Rickett, Leah Feltham, Kate Latham, Caroline Butler, Sarah Bennie, Lucy Brown, Rae Shirvington, Bethany Wood, Alice Latham, Mairead Loftus, Serena Nazareth, the Ebury sales team, Helen Crawford-White Nicky Gyopari and Julia Kellaway for helping me make this book the best it could possibly be.

I would also like to thank Sydney Rogers, Suzanne Quist, Gideon Weil and Judith Curr at HarperOne, my publishers in the USA.

My team at Tara Swart Inc. have supported me throughout the process of writing, and above and beyond. Thank you, Tracie Davis, Louise Malmstrom, Gillian Jay and Sara Devine.

It was my role as neuroscientist-in-residence at Corinthia Hotel, London that got me in touch with Joel, so thank you to everyone who worked with me there, especially Fiona Harris, Rica Rellon and Thomas Kochs.

I would like to thank Jules Chappell, Jen Stebbing, Flora Blackett-Ord and Johanna Pemberton for all their support, and the idea of the Corinthia residency that led to this book. Also to Matthew Wright for introducing me to Jules, and his deep support of me and my work.

Thank you to all my clients, colleagues and former patients for the richness of all the experiences in the book.

Finally, I would like to thank my friends and family for putting up with me for the duration of writing this book.

Index

Note from the Author

Thank you so much for reading *The Source*! I hope you feel that you have set your life on a new and exciting trajectory. What lies ahead will come from building on your awareness, actions and thinking in every area of your life. The steps in this book made a huge difference for me!

I love to hear from people who have tried these techniques. Often incredible changes and achievements have materialized in their lives from pictures on their action boards. If you are joining the group of people who celebrate the power of The Source, I would love to hear from you. Please stay in touch and share on Twitter or Instagram.

Twitter: @taraswart
Instagram: drtaraswart